Interpretability for Industry 4.0 : Statistical and Machine Learning Approaches

Antonio Lepore • Biagio Palumbo •
Jean-Michel Poggi

Editors

Interpretability for Industry 4.0 : Statistical and Machine Learning Approaches

 Springer

Editors
Antonio Lepore
University of Naples Federico II
Naples, Italy

Biagio Palumbo
University of Naples Federico II
Naples, Italy

Jean-Michel Poggi
University Paris-Saclay
Orsay, France

ISBN 978-3-031-12401-3   ISBN 978-3-031-12402-0   (eBook)
https://doi.org/10.1007/978-3-031-12402-0

This Springer imprint is published by the registered company Springer Nature Switzerland AG
The registered company address is: Gewerbestrasse 11, 6330 Cham, Switzerland

Preface

Interpretability is a key issue to develop insightful statistical and machine learning (ML) approaches in business and industry. This book aims to provide the readers with a compact, stimulating, and multifaceted introduction to this emerging promising topic.

The contents covered by the volume were stimulated by the ENBIS (European Network for Business and Industrial Statistics, https://enbis.org/) Workshop, Interpretability for Industry 4.0, which was held at the University of Naples Federico II Italy on July 12–13, 2021 https://conferences.enbis.org/event/8/ and offered real-world industrial motivations and deep methodological insights on the topic of interpretability. The workshop was divided into the following three pillars:

- Analyze and propose monitoring tools for additive manufacturing systems.
- Explore the connections between ML tools, sensitivity analysis, and rule-based systems.
- Exploit the contribution of generalized additive models for the development and visualization of interpretable statistical models.

Each half day, devoted to a specific pillar, ended with a roundtable providing a closing discussion challenging the different views of interpretability.

The book collects contributions issued from this workshop and accepted after a review process. It contains four chapters, the first one related to roundtables and three other chapters related to the three pillars mentioned above. Each chapter can be read independently with its own bibliography.

Chapter 1 introduces different views of interpretability in the context of Industry 4.0. It is organized in three different sections, after an introductory discussion about the concepts of explainability and interpretability of ML models. The chapter offers a philosophical discussion about the implications of ML interpretability for scientific and industrial studies and extends the concept in many directions, such as the generalizability of model outputs and implications for Industry 4.0 applications. The last section provides the reader with more specific materials and is dedicated to the connections between ML model interpretability and sensitivity analysis.

Chapter 2 discusses how the advent of artificial intelligence for manufacturing data mining poses new challenges to model interpretability in contrast with the concept of explainability. Starting from a general overview, the chapter focuses on examples of big data mining in additive manufacturing. A real case study focusing on spatter modeling for process optimization is discussed, where a solution based on robust functional analysis of variance is proposed.

Chapter 3 proposes a contribution to interpretability via random forests (RF) and is organized into two sections after an introduction about the need in different applications and domains for interpretable ML models and points out the requirements desired for interpretable methods. The chapter describes an original way to use RF to produce a compact set of rules and offers a thorough overview and analysis of permutation variable importance measures based on RF and describes new variants.

Chapter 4 formally introduces generalized additive models (GAMs) and flexible GAMs for location shape and scale (GAMLSS) as excellent models to achieve interpretability in the model building, as well as for communicating modelling results. Structural assumptions to avoid the curse of dimensionality in the modelling of the effect of a covariate vector on the distribution of a response variable are discussed. In particular, the additive assumption, on which GAMs rely, ensures scalability in the number of covariates and computational convenience in model fitting. The closing section of the chapter shows how to practically apply GAM and GAMLSS models via the `mgcv` and `mgcViz` R packages.

We are grateful to all the authors for their challenging perspectives on yet non-consolidated topics but highly relevant in supporting human decisions. We warmly thank anonymous referees for their conscientious reviews and Eva Hiripi from Springer-Verlag for supporting this project.

Naples, Italy   Antonio Lepore
Naples, Italy   Biagio Palumbo
Orsay, France   Jean-Michel Poggi
April 2022

Contents

Chapter 1
Different Views of Interpretability

Bertrand Iooss, Ron Kenett, and Piercesare Secchi

Abstract Interpretability, in the context of machine learning, means understanding the predictions made by the machine learning algorithm, with the aim to support human decisions based on them. In this view, interpretability can involve identifying the input features which drive the predictions. This chapter develops different issues and related methodologies of interpretability of machine learning models. Their implication for scientific and industrial studies are firstly developed. Then, the links between the generalizability of model outputs and interpretability are discussed. Finally, the deep connection between the settings of the machine learning interpretability and the ones of the model output sensitivity analysis is described.

1.1 Introduction

Machine Learning (ML) is one of the substantial branches of artificial intelligence technology and provides a large panel of algorithmic tools to learn from data (e.g., numerical data, images, sounds, texts). However, a severe drawback is that ML algorithms may provide predictions which turn out to be difficult to explain or interpret. From a general point of view, allowing an understandable explanation for any ML model output helps anybody (e.g., an operator, a decision-maker, a statistician or an analyst) to catch the underlying reasoning. Such a property may have positive consequences such as making the debugging process easier, helping for model improvement and acceptability of the tool. Therefore, industrial

B. Iooss (✉)
EDF R&D, Chatou and SINCLAIR AI Lab, Saclay, France
e-mail: bertrand.iooss@edf.fr

R. Kenett
KPA Group and Samuel Neaman Institute, Technion, Israel
e-mail: ron@kpa-group.com

P. Secchi
Department of Mathematics, Politecnico di Milano, Italy
e-mail: piercesare.secchi@polimi.it

deployment of these solutions requires tools together with a panel of best practices to perform explainable and interpretable ML. These two terms "explainability" and "interpretability" will be discussed, hereafter, in the context of ML.

There is a lack of consensus about rigorous definitions of explainability and interpretability of ML models. Indeed, these notions refer to profound cognitive processes related to social sciences and to their different fields of applications (e.g., medical sciences, law and justice, engineering) or scientific communities (e.g., natural language processing and computer vision). Some authors also invoke other fundamental concepts (see, e.g., completeness, fairness, intelligibility, comprehensibility, transparency) to build a proper definition of what "explainable AI" is and what it is intended for [5, 23]. In this chapter, we only focus on the ML interpretability as the property related to the ability of a ML model (or any element related to this model, i.e., inputs, outputs, predictions) to be associated with concepts held by a human being.

Interpretability, in the context of ML, means understanding the predictions made by the ML algorithm, with the aim to support human decisions based on them. In this view, interpretability can involve identifying the input features which drive the predictions. The goal of this chapter is to focus on different important issues and related methodologies of interpretability of ML models.

Firstly, the implication of ML interpretability for scientific and industrial studies are developed (Sect. 1.2). Then, the links between the generalizability of model outputs and interpretability are discussed, providing a high-level view with its implications to Industry 4.0 applications (Sect. 1.3). The approach presented combines an engineering perspective with empirical modeling and soft data in a blended hybrid view which integrates technical and non-technical perspectives. Finally, the deep connection between the settings of the ML interpretability and the ones of the model output sensitivity analysis is described (Sect. 1.4). This connection, which is still underdeveloped, offers rich perspective for cross-fertilizing techniques of both research fields [62].

1.2   Interpretability: In Praise of Transparent Models

The focus of ML is the design of algorithms that learn from a training data set how to associate an input to an output. The training data set must be massive since the learning curve of an ML algorithm increases very slowly with the size of the data set. The performance of a trained algorithm is typically evaluated on the task of prediction and validated with a hold-out data set. If future data are generated by the same population from which the training data set has been drawn, well trained ML algorithms are often excellent predictors.

Interpretability, in the context of ML, means understanding the predictions made by the algorithm, in order to support human decisions based on them. Interpretability involves identifying, for example with the tools of sensitivity analysis, the input features which drive the predictions. In this section we discuss if interpreting predictions, important as they are, is sufficient for science and for industry.

Quoting Carlo Rovelli's recent essay, *Helgoland* [64]: *"The goal of science is not that of making predictions. Science also aims at presenting an image of reality, a conceptual framework where to think about things. This is the ambition which made the scientific thinking successful. If predictions were the only goal of science, Copernicus would not have discovered anything different from Ptolemy: his astronomical predictions were not any better than those of Ptolemy. But Copernicus found the key for rethinking all and for better understanding it"*. This passage presents predictions as a partial objective indicating that interpretability should consider a wider scope.

A conceptual framework where to think about things is often required in business and industry. The broader question it therefore when and why, in science as well in business or in industry, do we use data? Briefly, we use data to answer three questions: what happened, what will happen, and what shall be done to make it happen.

1.2.1   What Happened?

This is the question tackled by exploratory and descriptive analyses that, starting from raw data, organize them, fuse different and heterogenous sources impinging upon the same population, sort them out deciding about the relevant and the irrelevant, clean and transform data, graphically represent and summarize the information sufficient for the goal of the analysis, already driving it toward certain hypotheses and conjectures. For being effective, and not mystifying, an exploratory analysis must be open, totally transparent and highly dialectical. Through it the different stakeholders, who promoted the final questions moving the problem tackled with data, should be guided to better formalizations and prioritizations. This is an intensely Human Intelligence (HI) stage, where the data scientists are, explicitly or implicitly, guided by models. For instance, when they discriminate between the features of interest and those that will in fact not be measured and recorded, or when they choose the proper mathematical representation for data. Are the atoms of the analysis time series or functional data? How should time dependence be captured within each datum, explicitly through the autocorrelation function or implicitly by imposing certain smoothness and regularity to their functional representation? Researchers are in fact usually called to decide the specific mathematical space to embed data at hand and thus the geometry that allows for distinct projections and dimensionality reductions, the main mathematical tools for compressing and transferring the sufficient and relevant information.

Picture Galileo entering the Pisa Cathedral and observing the swinging chandelier. Did he observe and record the temperature and humidity of the air in the room in previous days and months, the number of people attending mass, their gender or their age, the phase of the moon, the hour and the day of the year, etc.? In the big data era he might have, but at the end of the sixteenth century [52] Galileo decided to focus only on the periods of the pendulum, the amplitude of the swing, the length of the rod and the mass of the bob. All other data were discarded and considered in advance as non-influential, even before measuring them. Surely this must have been decided based on intuition, a model which was forming in his mind, about the not yet formalized isochronism law of the pendulum, which in the following decades elected it as the disruptive new technological device for timekeeping.

1.2.2 What Will Happen?

This is the stage when we want to make predictions. We use training data, validation data and test data, to build predictor machines and evaluate their performances. These can be transparent models, like generalized additive models mixing endogenous and exogenous variables, opaque models, like random forests, or black boxes, like deep neural networks. They can be subjected to natural interpretability—at least for the educated data scientist—or they might be inaccessible and require the tools of sensitivity analysis to elicit the contribution of the input features, and their interactions, to generate the final output. Uncertainty quantification is here a must. Different approaches have indeed been cleverly elaborated in the past centuries for the purpose—frequentist inference, Fisherian inference, Bayesian inference, Monte Carlo methods, bootstrap, cross-validation. The very concept of uncertainty has been fragmented many times—aleatoric uncertainty, epistemic uncertainty, forward uncertainty propagation, inverse uncertainty quantification, etc.

1.2.3 What Shall be Done to Make It Happen?

This is the realm of prescriptive analysis and experimental design. Assuming some input data provided by idealized scenarios and given the predictions offered by the models, what actions should be taken in order to generate the desired effects, with a certain degree of certainty. Once more, quantification of uncertainty is a must. But how should sensible scenarios be built? Can they be totally ignorant of the past as captured and represented by the sufficient summaries provided by the exploratory and descriptive analyses? A domain-based HI, transparency and a dialectical perspective are the effective trading tools. Indeed, here again the transparent models—in science, business and industry—are the empowering tools for sharing empirical and «experiential»knowledge, across different teams and units, across generations of scientists, engineers and experts.

In fact, the question is at the basis of experimental design and significance testing. Quoting Fisher [21]: *"We may say that a phenomenon is experimentally demonstrable when we know how to conduct an experiment which will rarely fail to give us statistically significant results."*

1.2.4 Patterns and Models

A model is not a magic box. Its value resides in its power to amplify the human thought. This happens if the model is able to represent the interactions and the dependencies among the variables that both the stakeholders and the data scientist believe are describing the system under scrutiny. A model allows for sensible decision making through the action on independent input variables; a model provides scope for simulation and manipulation of the system under scrutiny.

Transparent models, opaque models and black boxes permit interpretability which enable an incremental upgrade of the human knowledge. This requires more than an automated and theory-free data analysis. Contrary to Chris Anderson's dictum on *Wired* [2], "correlation is NOT enough".

Beside predictions, the other selling point of ML algorithms is their ability to find patterns in massive data without the intermediation of a theory, without moving through the slow process of identifying a formal reference system within which questions could be asked and hypothesis could be stated.

As argued by James McAllister [47], any *"empirical data set can be decomposed into any one of all conceivable patterns and an associated noise term."* Hence, only two options seem admissible. Either we assume the ability of the ML algorithm to discern among patterns those that are indicative of real structures of the world and those that are not. Or, in the absence of an automatic criterion implementing this discrimination, we are forced to follow McAllister's argument and *"deny that any such ontologically significant distinction between patterns can be drawn, to admit that all patterns exhibited in empirical data sets correspond to structures in the world, and then to consider the meaning and implications of this claim."*

Without entering any further into this intriguing philosophical debate, let us notice that if one is looking for patterns, the analysis of a big data set formed by the decimal digits in the expansion of Pi, the ratio of any circle's circumference to its diameter, has the potential to fill one's life. Pi is an irrational and transcendent number, whose approximate representation through a decimal expansion is often used as a test for evaluating the power of new supercomputers; the last record has been broken in August 2021, when Pi has been accurately approximated to 62.8 trillion decimal places [73]. By using the MyPiday [75] search engine, one of us found among the digits of Pi his birthday, that of his wife, the day they married and the birthdays of their children... a pattern of a certain relevance, at least to him. An (unproven) conjecture states in fact that Pi is a normal number (see, e.g., Arndt and Haenel [4]), which would also imply that "every finite string of numbers eventually occurs in Pi". Structures representable by finite string of numbers should be able to accommodate the answers to all problems business and industry might want a data scientist to solve, and yet we would not consider as reasonable and practical to search the digits of Pi for finding them. The problem being that, without knowing it in advance, we will not be able to recognize a relevant and correct answer if we met it among the decimal digits of Pi, although we know that is there...

This sounds as an anti-climax for the believers in the automatic heuristic power of the ML algorithms, but it should not. The problem is with the "automatic" qualification. We could indeed use ML to explore data in search for patterns if we admit that our search is driven by our intention, by the objectives of our endeavor. It is the intention driven by the goal of the analysis and framed within a theory, explicitly or implicitly formalized, which generates the relevant conjectures and the hypothesis the data could be challenged with; it is intention which puts the data scientist in the position to decouple the patterns within the data deemed to be relevant from those that are not. Eliciting this intentionality is more easily achieved with a transparent model, where interpretation is—to paraphrase Karl Pearson [70]—"on the table", but could in principle also be obtained with the tools of Artificial Intelligence, through a stronger and still unusual effort and the development of new mathematical—and transparent—perspectives.

1.3 Generalizability and Interpretability with Industry 4.0 Implications

In this section we focus on the process of moving from numbers to data, to information and insights [31]. In the information quality framework, this is called "generalizability" an expanded form of "interpretability" [36]. The section covers interpretable artificial intelligence (AI), wide angle of statistical generalizability.

1.3.1 Introduction to Interpretable AI

Artificial Intelligence (AI) has focused on predictive analytics with success reflected by sophisticated black box models. In recent years, the need to interpret and explain the factors affecting analytic predictions has risen. To achieve this, various methods have been proposed to help users interpret the predictions of complex models. Lundberg and Lee [44] introduce SHAP (SHapley Additive exPlanations), a unified framework for interpreting predictions. SHAP assigns to each feature in the model an importance value for a particular prediction. It includes the identification of additive feature importance measures and theoretical results showing there is a unique SHAP solution with a set of desirable properties.

Local interpretable model-agnostic explanations (LIME) is a local surrogate interpretable model used to explain individual predictions of black box ML models [49, 63]. Surrogate models approximate the outputs of a black box model [13, 22]. LIME is based on local surrogate models used to explain individual predictions. In a first step, LIME uses the black box model to get model predictions, ignoring the training data. The objective is then to understand why the ML model gives a certain prediction. LIME generates a new dataset using perturbed samples and then generates ML model predictions. On this new dataset, LIME trains a simple (interpretable) model, weighted by the proximity of the sampled instances to the instance of interest. The new model is not supposed to be a global approximation of the ML model, but only a good local approximation. Mathematically, local surrogate models with interpretability constraint, is expressed as:

$$\text{explanation}(x) = \text{argmin}_{g \in G} L(f, g, \pi_x) + \Omega(g).$$

The explanation function for instance x is the model g that minimizes loss L which measures the distance between the explanation model and the prediction of the original model f (e.g., an xgboost model). The model complexity $\Omega(g)$ is kept as low as possible, g is the family of possible explanations, for example all possible linear regression models. The proximity measure, π_x, defines the neighborhood size around instance x that we consider for the explanation.

1.3.2   A Wide Angle Perspective of Generalizability

In discussing interpretability and generalizability we can consider both hard and soft data. Data that is "soft," is based on a person's experience, impressions, and understanding. Soft data stands in contrast to "hard data," particularly that which is digitized. While methods like SHAP and LIME are based on hard data, interpretability must not discount soft data. In fact, good science often combines the two.

Figure 1.1 summarizes several generalization methods. In generalizing findings from models, we want to make valid inferences and predictions from the data about other areas of interest to decision-makers. Statistical inference is making claims about the stochastic behavior or population frame we observed with the collected data. We identify several distinct methods for generalizability [35, 36]:

- "The Laws of nature" refers to laws and models that allow one to extrapolate, under assumptions;
- "Statistical generalizability" refers to making inference from a sample (of hard data) to a target population;
- "Domain-specific generalizability" refers to applying domain knowledge, not fully supported by the hard data, to other circumstances, such as the future or different populations; and
- "Intuition" refers to people's ability to reason from the data in ways that cannot be fully explained. Science in general and data science in particular often discounts intuition. But it is undeniably true that some decision makers have unerring intuition. And at the least it is necessary because all decisions are made in the face of uncertainty.

A reliable form of generalization involves the laws of nature. These include Conservation of Mass, Conservation of Energy, Conservation of Momentum, Newton

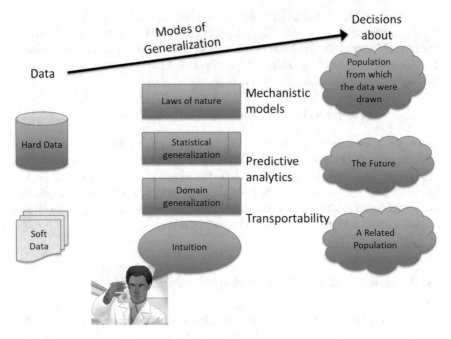

Fig. 1.1 Modes of generalization (or interpretability)

Laws, Principle of Least Action, the Laws of Thermodynamics and Maxwell's equations. Sometimes these are called mechanistic models of modes of action. These laws started as empirical laws that were embraced as laws of nature and have withstood the test of time. They have been verified time and again and today. We invoke them without the need for data, only knowledge of physics, chemistry, biology or other scientific disciplines.

Transfer learning is yet another type of generalizability. A mathematical and geometrical model of transfer learning in neural networks training was proposed in Bozinovski and Fulgosi [14]. Algorithms are available for transfer learning in Markov logic networks [48] and Bayesian networks [32, 53], See also Yang et al. [76].

Mathematics offers a unique context for such interpretability or generalization. Paul Erdös used to talk about "The Book", in which God maintains the perfect proofs of mathematical theorems [1]. Paraphrasing Galileo once again, the laws of nature build on The Book.

1.3.3 Statistical Generalizability

Now consider statistical generalizability. In making inference about a population parameter, from a sample, statistical generalizability and sampling bias are the focus a key question of interest is "What population does the sample represent?" [18, 60]. In contrast, for predicting the values of new observations, the question is whether the analysis captures associations in the training data (i.e., the data used in model building) that generalize to the to-be-predicted situations, or out of sample conditions. Control charts present a good example. Assuming the process remains stable, we expect performance to vary within the upper and lower control limits [38].

Statistical generalizability is commonly evaluated using measures of sampling bias and goodness of fit. In contrast, scientific generalizability, used for predicting new observations, is often evaluated by the accuracy of prediction of a hold-out set from the to-be-predicted population. This assessment is a crucial protection against overfitting, which occurs when your model fits previously collected data perfectly but does very poorly with new data [69].

Randomization is a key approach in interventions that enables statistical generalization. As well as guarding against unknown biases, it provides the mathematical foundations that support calculation and interpretation of p-values. However, clinical trials with randomized allocation of patients to treatment or placebo, , may be subject to "sample selection-bias," since participation in a randomized trial cannot be mandated. Sample patients may consist of volunteers who respond to financial and medical incentives, leading to a distribution of outcomes in the study that differ substantially from the distribution of outcomes more generally. This sample selection bias is a major impediment in both the health and social sciences [25]. In web application development and many software testing initiatives, one uses A/B testing where users are randomly presented with version A or version B. The user experience such as click conversation statistics in versions A and B are compared. Sample bias can also affect A/B testing [38].

"Transportability" is another way to generalize. Transportability is defined as a transfer of causal effects learned in experimental studies to a new population, where only observational studies can be conducted [57]. In a study on urban social interactions, Pearl and Bareinboim [58, 59] used transportability to predict results in New York City, based on a study conducted in Los Angeles, accounting for differences in the social landscape between New York City and Los Angeles.

Another example of generalization, in the context of personal ability testing, is specific objectivity [61]. This testing is known as item response testing (IRT). Specific objectivity is a theoretical state achieved if responses to a questionnaire used to compare levels of tested individuals are generalizable. In other words, the purpose is to learn from a set of questions on the population frame of outcomes.

Online auction studies provide another example of the importance of generalization. A study of the effect of reserve price on eBay auction final price is reported in Katkar and Reiley [30]. The authors designed an experiment that produced a representative sample of recorded auctions. Their focus was on statistical generalization of the impact of reserve price on auction outcomes. In contrast, the study by Wang et al. (2008) [72] forecasts the results of new auctions. In that study, the authors, evaluated predictive accuracy using a hold-out set as opposed to standard errors and sampling bias considered by the first study. A third study, on consumer surplus in eBay, dealt with statistical generalizability by inferring from a sample to all eBay auctions. Because the sample was not drawn randomly from the population, Bapna et al. [7] compare their sample with a randomly drawn sample.

Domain-based (or scientific) expertise allows findings from specific data to be applied more generally [36]. Thus, a marketing manager might base his decisions on how to run a marketing campaign in location A, using a market study conducted in location B. He has no data on A, but his experience (soft data) tells him how to adopt the conditions in B to what is required in A. Similarly, a software development manager, in the face of a limited testing budget, might decide to release a version with minimal testing because its functionality is basic and the person who developed it has a good record. In other cases, he might decide to significantly increase the testing effort. Such decisions are often made without formal data analyses. The approach has benefits (i.e., speed), but carries risks that decision makers should bear in mind.

Industry 4.0 requires generalization and interpretability to derive gains in productivity and enhanced flexibility. A manager of a production line with several CNC metal work robots needs to assign specific work orders to specific systems. Past experience on the behavior of these systems, and engineering consideration of the parts to be manufactured, is used in such assignments [29, 39].

Simulations, in general, aim at generalization and interpretability by modeling in detail physical phenomena using finite element methods or by invoking emulators [37]. In that context, digital twins are now used to provide monitoring, diagnostic, prognostic and prescriptive analytic capabilities [33, 34]. In contrast to off-line simulators, digital twins are fed by online sensors and generalize physical assets in the digital space. Moreover, the diagnostic capabilities delivered by digital twins provide unique interpretability of events identified through monitoring.

1.4 Connections Between Interpretability in Machine Learning and Sensitivity Analysis of Model Outputs

This section[1] develops the connections between the understanding of ML models (called here the "ML interpretability") and the topic of sensitivity analysis (SA) of model outputs. The first subsection illustrates these connections by schematizing their methodologies, while the two following subsections summarize and parallelize their main settings. A last subsection provides a brief review of specific SA methods that have been transposed to serve the ML interpretability.

---

[1] This section has been written with the support of Vincent Thouvenot (Thales and SINCLAIR AI Lab, France) and Vincent Chabridon (EDF R&D and SINCLAIR AI Lab, France).

1.4.1 Machine Learning and Uncertainty Quantification

Interpretability is linked to the ability, for a human mind, to understand represen-
tations of the ML model such as resulting predictions and associated decisions.
Regarding this point, one needs to recall that ML models are intensively used
to make predictions on output quantities of interest. These quantities are then
used to make decisions (e.g., regarding safety criteria or economic efficiency and
profitability ones). This leads to connect interpretability in ML with the topic of SA
of model outputs [19, 65]. SA is most often included in the uncertainty quantification
process [67] of a numerical simulation model (e.g., a partial differential equations-
based computer model which represents a physical phenomenon).

Figure 1.2 gives an overview of the uncertainty quantification methodology,
developed for example in Baudin et al. [8]. Such a process is often carried out
to consider and better understand sources of imprecision when using numerical
modeling. The different steps involved are clearly highlighted. In Step A, the model,
its inputs, its outputs and the quantity of interest of the study are specified. In Step
B, the uncertainty model for the inputs (e.g., their joint probability distribution) is
built, possibly (Step B') by calibration and validation on observed data. In Step C
the uncertainties are propagated through the model. Finally, in Step C', the SA and
the robustness analysis are performed. In this context, SA involves studying how
the uncertainty in the output of the model can be apportioned to different sources of
uncertainty in its inputs. For instance, SA may be used to determine the inputs that

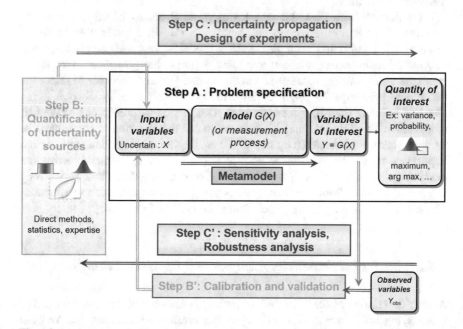

Fig. 1.2 Methodological scheme of the uncertainty management of computer models

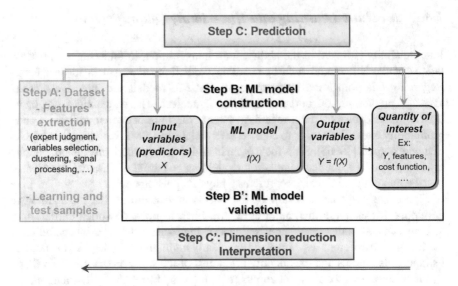

Fig. 1.3 Methodological scheme of the machine learning process

most contribute to some output behavior, detect noninfluential inputs, and uncover interaction effects within a model. SA can also be used to rank the influence of inputs as well as to calibrate, verify, understand, simplify the model and then guide future research efforts.

As a parallel with Figs. 1.2 and 1.3 gives an overview of the ML process, focusing on supervised learning, i.e., building a statistical model from a set of labeled (i.e., input-output) samples. Such a model is then used to predict, for a new set of input values, the corresponding output value [26]. Step A concerns the extraction (from the dataset) of the inputs and features (in order to select the ML model predictors), as well as the learning and test samples. Step B is the phase of building the ML model, denoted $f(x)$. In step C, predictions of the ML model are made while ML model interpretation and the detection of non-influential inputs/features are considered by step C'. Thus, the variable selection methods and importance measures, that can be used in step C', share the same objective and mathematical framework as SA techniques. For example, there is a very strong connection between Sobol' sensitivity measures [68] and importance score methods obtained when using resampling techniques like random forests [10].

1.4.2 Basics on Sensitivity Analysis and Its Main Settings

A clear description of the objectives of a study is essential before performing SA. Considering both a methodological and an engineering point of view, Da Veiga et al. [19] contemplates the following four important SA settings:

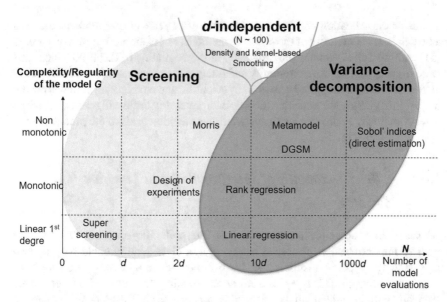

Fig. 1.4 Basic grid classification of sensitivity analysis methods (*d* corresponds to the number of model inputs)

1. *Model exploration* which aims to understand the behavior of a model by trying to investigate the input-output relationship, e.g., starting with an initial visual analysis of the input-output sample.
2. *Factor fixing* which aims to reduce the number of uncertain inputs by setting unimportant factors as constants. Unimportant factors are those that, if set to any particular value, do not lead to a significant reduction in output uncertainty.
3. *Factor prioritization* which aims to identify the most important factors. For instance, in the case of independent inputs, the most important one is that which, if set to a specific constant, would lead to the largest reduction in uncertainty in the quantity of interest.
4. *Input distribution robustness* which aims to analyze variation in the quantity of interest with respect to uncertainty in inputs' distributions.

Then, a large amount of SA techniques has been developed [19, 65]. Figure 1.4 summarizes the most famous ones in a simple classification grid following two axes: the abscissa is about the size of the input-output sample required to use the method and the ordinate asks for the underlying hypothesis made on the structure of the model. One may also distinguish two classes of methods: the qualitative ones (screening) aiming at detecting non-influential inputs and the quantitative ones (variance decomposition) aiming at quantifying the effects of the inputs on the output variance. Of course, more refined grids have been developed to provide guidance to the practitioners about the use of the different estimation techniques associated to the different SA methods [19].

Such a classification grid has been obtained after years of development and then a certain level of maturity of the different methods. For example, the Morris method [51] and the Sobol' indices [68] appeared at the beginning of the 1990s, then they were discussed in the SA community and integrated in various software. They have also been intensively applied across many scientific and engineering domains during the last thirty years. We hope that the ML interpretability topic will reach a sufficient level of maturity to provide some adequate classification grids to the practitioners.

1.4.3 A Brief Taxonomy of Interpretability in Machine Learning

As stated in Molnar [49], a taxonomy of interpretability methods can be built with various criteria, as described below. A first classification could be proposed following the type of results given by the method (e.g., visualization tools or summary statistics). This is useful from a user-friendly perspective. However, it does not really help to understand the underlying similarities and differences between the methods.

The literature on the subject often distinguishes intrinsically interpretable models and model interpretability techniques. In the first case, models are called "transparent", while in the second case, the techniques are referred to as "post-hoc interpretability" [5, 49]. By post-hoc techniques, one considers that a supplementary layer of statistical quantities (e.g., sensitivity indices, feature importance measures) is required to achieve a proper interpretation of the model results and a full understanding of the model behavior. As an example, linear models (such as linear and logistic regressions) are often assumed to be the prototype of intrinsic transparent models. However, even for this simple class of models, direct interpretability might be tricky for many reasons (as, e.g., strong dependencies between inputs). Thus, post-hoc interpretability seems to be inevitable and suitable in many real-world applications.

Another dichotomy is between model-specific and model-agnostic methods [49]. The first ones are specific to a given class of models while the other ones can be used on any ML model which has been trained on data. The idea is to characterize the genericity of the interpretability methods.

Finally, a strong distinction exists between global and local interpretability. On the one hand, a global interpretation aims at explaining the behavior of a model throughout his validity domain (i.e. for all the population of observations), and to identify the features that are the most influential, globally speaking. On the other hand, a local interpretation tries to catch the effect of a specific feature on one specific (or a group) of model predictions [5, 43, 49]. Again, these notions show strong links with similar (but different) notions of "local" and "global" SA techniques [19].

Then, by making a parallel view with the SA settings explained in Sect. 1.4.2, four different objectives can be distinguished in the context of ML interpretability (see also the settings defined in Bénard [9]):

1. *Visualization* of the relation between the label and the features. This can be made via the use of partial dependence plots, individual conditional expectation, accumulated local effects, etc.
2. *Features identification*. The objective of this setting is to find a small number of inputs that reach a maximized accuracy of the ML model. Identifying features in the data involves the use of statistical techniques such as correlation measures (Pearson coefficient, Spearman's rank correlation coefficient, Kendall's tau, copula, Hoeffding's D) and kernel-based metrics. Even if with different objective, it corresponds to the factor fixing setting in SA and mainly concerns screening techniques.
3. *Measures of importance* of explanatory variables. It consists in measuring the impact on the outputs of the inputs/features. It corresponds to the factor prioritization setting in SA, where the goal is to detect and rank influential inputs to explain the model to experts. One can be interested in two types of impact. The first one, which is global, consists in explaining how a feature or a set of features impact the output distribution. The second one, which is local, consists in explaining how the features impact the model's output for one instance.
4. *Robustness* of the decision boundary. This last setting concerns two topics: (a) the counterfactual examples (i.e., explaining the prediction related to one individual by another close individual with an opposite target prediction); (b) how the output label changes when the distribution of one input or group of inputs change (perturbations of the data distribution). It corresponds to the input distribution robustness setting in SA.

1.4.4 A Review of Sensitivity Analysis Powered Interpretability Methods

Many researchers and practitioners judge that current ML interpretability methods are most often approximative and not clearly defined (see, e.g., Molnar et al. [50]). Thus, their operational diffusion can be difficult, and their results are often subject to caution. In contrast, many importance measures have been developed with strong theoretical bases in SA, and then finely studied from numerical and applied points of view. However, despite the bright connections between SA and ML interpretability, most of the researchers from the ML community seem to ignore the SA scientific field. This can be illustrated, for instance, by the recent work of Covert et al. [17] which develops a unified framework for model explanation metrics, without referring to any SA-based concepts nor SA literature. The reverse is not true as proved by the recent collective paper of Razavi et al. [62]. Coming from the SA community, it advocates that the cross-fertilization between SA and ML is a wide avenue of upcoming research. In this context, first works have recently shown how to take advantage of the SA tools to robustify ML global interpretability. These have been mainly done for the settings of features identification and importance measures, where the variance-based importance measures of SA have proven to be prolific.

For features ranking in ML, importance score methods associated to permutation and resampling techniques [15] are widely used, for example via the random forests ML model [3]. As explained and reviewed in the latter reference, the permutation-based importance measures correspond to the total Sobol' indices [27] which are one of the most widely used tool in SA. Therefore, the deep theoretical works and the plethora of estimation methods developed for Sobol' indices (see [19] for a review) help to understand the properties of the different definitions of the permutation-based importance measures [10]. In addition, these theoretical works on SA techniques can help to provide new importance measures that are consistent in case of dependent inputs, as also done in Bénard et al. [10] via the random forests technique.

Several SA metrics are devoted to identifying the most influential features learned by a black-box model (e.g., a ML model), as well as to detect interactions between features. These SA metrics can be used to identify optimal levels of ML model structural complexity, for instance to help designing deep neural networks. Lauret et al. [40] and [42] have used SA tools to prune redundant neurons in artificial neural networks model. Fel et al. [20] have shown how Sobol' indices can capture high-order interactions between image regions and their contributions to a neural network. Thanks to SA-based efficient computation schemes of total Sobol' indices, Owen and Hoyt [56] and Hahn et al. [24] use the notion of mean dimension (which represents the mean interaction degree between inputs that acts on the output of a model) to characterize the internal structure of complex modern neural network architectures. The mean dimension allows to summarize the extent to which the ML model is dominated by high or low order interactions. This tool is then useful to analyze the neural networks architectures, associated to accuracy metrics. The mean dimension also allows a deep analysis about which layers of the neural network occur. Finally, Novello et al. [54] propose the use of the Hilbert-Schmidt independence criterion analysis (developed in a ML features identification context and incorporated several years ago as a SA tool; see [19]), for the analysis, interpretability, and optimization of hyperparameters of deep neural networks.

For the importance measure setting, the Shapley effects [55] and the SHAP metric [44] are both based on the Shapley values concepts [66] and developed during the same period for SA and ML interpretability, respectively. In SA, the main advantage of Shapley effects over Sobol' indices lies on the correct treatment of the dependent inputs' case (see [19] for a review), often encountered in practice. For the same reason, SHAP is intensively used in ML interpretability but remains a local metric, contrarily to the Shapley effects which are global, and explains each individual prediction (see Sect. 1.3.2). To understand the impact of one feature across a whole dataset when building a ML model, global importance measures are required. [16] introduced SAGE, for Shapley Additive Global Importance, to propose a general view of global feature importance for different types of learning models and loss functions. Shapley effects correspond to a notable specific case of SAGE. [74] proposed a computationally efficient procedure for estimating the so-called Shapley population variable importance measure, which is actually a Shapley effect with any value function. Bénard et al. [11] have recently introduced SHAFF (SHApley eFfects via random Forests), a fast and accurate estimator of Shapley effects, based on the use of the random forests technique.

Note that ML model interpretability opens the door to model fairness. Fairness in ML consists in making sure a decision is not based on the so-called "protected attributes" [71], for instance to avoid human discrimination (e.g., gender, race... for a bank loan). A recent work [12] has shown that fairness can be seen as a special framework of SA and have linked SA metrics to fairness metrics. Mase et al. [45, 46] have also developed a Shapley values-based method (called cohort Shapley values) to evaluate the impact of learning a protected attribute, which can be external to the ML model, on the ML model predictions.

Finally, the robustness on the decision boundary setting has been recently addressed in Bachoc et al. [6] by using another SA method, called perturbed law-based sensitivity indices [28, 41]. The idea is to evaluate the new ML model predictions after a perturbation of a moment (e.g., the mean) of the data distribution (typically one feature at a time) by a small stress. No new data is needed as the stress is based on a statistical re-weighting scheme called entropic variable projection. The robustness metric is then a curve that gives the deviation of a scalar quantity of interest (for example the average prediction or a misclassification rate) as a function of the stress amplitude.

References

1. Aigner M, Ziegler G (2000) Proofs from THE BOOK. Springer-Verlag, Berlin
2. Anderson C (2008) The end of theory: the data deluge makes the scientific method obsolete. Wired magazine 16(7):16–07
3. Antoniadis A, Lambert-Lacroix S, Poggi JM (2021) Random forests for global sensitivity analysis: a selective review. Reliabil Eng Syst Safe 206:107312
4. Arndt J, Haenel C (2006) Pi unleashed. Springer, Berlin
5. Arrieta AB, Díaz-Rodríguez N, Ser JD, Bennetot A, Tabik S, Barbado A, Garcia S, Gil-Lopez S, Molina D, Benjamins R, Chatila R, Herrera F (2020) Explainable artificial intelligence (XAI): concepts, taxonomies, opportunities and challenges toward responsible AI. Inf Fus 58:82–115
6. Bachoc F, Gamboa F, Halford M, Loubes JM, Risser L (2020) Explaining machine learning models using entropic variable projection. Preprint. arXiv:181007924v5
7. Bapna R, Jank W, Shmueli G (2008) Consumer surplus in online auctions. Inf Syst Res 19(4):400–416
8. Baudin M, Dutfoy A, Iooss B, Popelin A (2017) Openturns: an industrial software for uncertainty quantification in simulation. In: R G, D H, H O (eds) Handbook of uncertainty quantification. Springer, New York
9. Bénard C (2021) Random forests and interpretability of learning algorithms. PhD thesis, Sorbonne Université

10. Bénard C, Da Veiga S, Scornet E (2022) MDA for random forests: inconsistency, and a practical solution via the sobol-MDA. Biometrika
11. Bénard C, Biau G, Da Veiga S, Scornet E (2022) Shaff: fast and consistent shapley effect estimates via random forests. In: Proceedings of the 25th international conference on artificial intelligence and statistics, virtual
12. Bénesse C, Gamboa F, Loubes JM, Boissin T (2022) Fairness seen as global sensitivity analysis. Machine Learning. https://doi.org/10.1007/s10994-022-06202-y
13. Box GE, Draper NR (1987) Empirical model-building and response surfaces. Wiley, New York, NY
14. Bozinovski S, Fulgosi A (1976) The influence of pattern similarity and transfer learning upon the training of a base perceptron B2. In: Proceedings of Symposium Informatica, pp 3–121
15. Breiman L (2001) Random forests. Machine Learning 45(1):5–32
16. Covert I, Lundberg SM, Lee SI (2020) Understanding global feature contributions with additive importance measures. Adv Neural Inf Process Syst 33:17212–17223
17. Covert I, Lundberg S, Lee SI (2021) Explaining by removing: a unified framework for model explanation. J Mach Learn Res 22(209):1–90
18. Cox DR, Kartsonaki C, Keogh RH (2020) Statistical science: some current challenges. Harvard Data Sci Rev 2(3). https://doi.org/10.1162/99608f92.a6699bda
19. Da Veiga S, Gamboa F, Iooss B, Prieur C (2021) Basics and trends in sensitivity analysis: theory and practice in R. SIAM
20. Fel T, Cadène R, Chalvidal M, Cord M, Vigouroux D, Serre T (2021) Look at the variance! efficient black-box explanations with sobol-based sensitivity analysis. In: Advances in neural information processing systems (NeurIPS) 34
21. Fisher RA (1935) The design of experiments. Edinburgh: Macmillan Pub Co
22. Forrester A, Sobester A, , Keane A (2008) Engineering design via surrogate modelling: a practical guide. Wiley, New York
23. Gilpin LH, Bau D, Yuan BZ, Bajwa A, Specter M, Kagal L (2018) Explaining explanations: An overview of interpretability of machine learning. In: Proceedings of the 2018 IEEE 5th international conference on data science and advanced analytics (DSAA), Turin
24. Hahn R, Feinauer C, Borgonovo E (2022) The Mean Dimension of Neural Networks - What causes the interaction effects? Preprint. arXiv:2207.04890
25. Hartman E, Grieve R, Ramsahai R, Sekhon JS (2015) From sample average treatment effect to population average treatment effect on the treated: combining experimental with observational studies to estimate population treatment effects. J R Stat Soc Ser A (Stat Soc) 178(3):757–778
26. Hastie T, Tibshirani R, Friedman JH, Friedman JH (2009) The elements of statistical learning: data mining, inference, and prediction, vol 2. Springer, New York
27. Homma T, Saltelli A (1996) Importance measures in global sensitivity analysis of nonlinear models. Reliab Eng Syst Safe 52(1):1–17
28. Iooss B, Vergès V, Larget V (2021) BEPU robustness analysis via perturbed law-based sensitivity indices. Proc Inst Mech Eng Part O: J Risk Reliabil. https://doi.org/10.1177/1748006X211036569
29. Kang S, Jin R, Deng X, Kenett RS (2021) Challenges of modeling and analysis in cyber-manufacturing: a review from a machine learning and computation perspective. J Intell Manuf 2021:1–14
30. Katkar R, Reiley DH (2006) Public versus secret reserve prices in eBay auctions: results from a pokémon field experiment. Adv Econ Anal Pol 6(2):Article 7
31. Kenett RS (2008) From data to information to knowledge. Six Sigma Forum Mag 8(1):32–33
32. Kenett RS (2019) Applications of Bayesian networks. Trans Mach Learn Data Mining 12(2):33–54
33. Kenett RS, Bortman J (2021) The digital twin in industry 4.0: a wide-angle perspective. Qual Reliab Eng Int 21(23):7830
34. Kenett RS, Coleman S (2021) Data and the fourth industrial revolution. Significance 18(3):8–9
35. Kenett RS, Redman TC (2019) The real work of data science: Turning data into information, better decisions, and stronger organizations. Wiley, Hoboken

36. Kenett RS, Shmueli G (2016) Information quality: the potential of data and analytics to generate knowledge. Wiley, Hoboken
37. Kenett RS, Vicario G (2021) Challenges and opportunities in simulations and computer experiments in industrial statistics: an industry 4.0 perspective. Adv Theor Simul 4(2):1–15
38. Kenett RS, Zacks S (2021) Modern industrial statistics: With applications in R, MINITAB, and JMP. Wiley, Hoboken
39. Kenett RS, Swarz RS, Zonnenshain A (2019) Systems engineering in the fourth industrial revolution: big data, novel technologies, and modern systems engineering. Wiley, Hoboken
40. Lauret P, Fock E, Mara T (2006) A node pruning algorithm based on a fourier amplitude sensitivity test method. IEEE Trans Neural Netw 17(2):273–293
41. Lemaître P, Sergienko E, Arnaud A, Bousquet N, Gamboa F, Iooss B (2015) Density modification-based reliability sensitivity analysis. J Stat Comput Simul 85(6):1200–1223
42. Li B, Chen C (2018) First-order sensitivity analysis for hidden neuron selection in layer-wise training of networks. Neural Process Lett 48(2):1105–1121
43. Longo L, Goebel R, Lecue F, Kieseberg P, Holzinger A (2020) Explainable artificial intelligence: concepts, applications, research challenges and visions. In: Holzinger A, Kieseberg P, Tjoa AM, Weippl E (eds) Machine learning and knowledge extraction. CD-MAKE 2020, vol 12279. Springer International Publishing, Cham, pp 1–16
44. Lundberg SM, Lee SI (2017) A unified approach to interpreting model predictions. In: Guyon I, Luxburg UV, Bengio S, Wallach H, Fergus R, Vishwanathan S, Garnett R (eds) Advances in neural information processing systems, vol 30. Curran Associates, Inc., Red Hook. https://proceedings.neurips.cc/paper/2017/file/8a20a8621978632d76c43dfd28b67767-Paper.pdf
45. Mase M, Owen AB, Seiler B (2019) Explaining black box decisions by shapley cohort refinement. CoRR. http://arxiv.org/abs/1911.00467
46. Mase M, Owen AB, Seiler BB (2021) Cohort shapley value for algorithmic fairness. CoRR. https://arxiv.org/abs/2105.07168
47. McAllister JW (2011) What do patterns in empirical data tell us about the structure of the world? Synthese 182(1):73–87
48. Mihalkova L, Huynh T, Mooney RJ (2007) Mapping and revising markov logic networks for transfer learning. In: Proceedings of the 22nd national conference on artificial intelligence - volume 1, AAAI'07. AAAI Press, Palo Alto, pp 608–614
49. Molnar C (2022) Interpretable machine learning: A guide for making black box models explainable. https://christophm.github.io/interpretable-ml-book/
50. Molnar C, Casalicchio G, Bischl B (2020) Interpretable machine learning – a brief history, state-of-the-art and challenges. In: ECML PKDD 2020 workshops. Springer International Publishing, Cham, pp 417–431. https://doi.org/10.1007/978-3-030-65965-3_28
51. Morris MD (1991) Factorial sampling plans for preliminary computational experiments. Technometrics 33(2):161–174
52. Murdin P (2008) Full meridian of glory: perilous adventures in the competition to measure the Earth. Springer, New York
53. Niculescu-Mizil A, Caruana R (2007) Inductive transfer for bayesian network structure learning. In: Meila M, Shen X (eds) Proceedings of the eleventh international conference on artificial intelligence and statistics, PMLR, San Juan, Puerto Rico. Proceedings of machine learning research, vol 2, pp 339–346. https://proceedings.mlr.press/v2/niculescu-mizil07a.html
54. Novello P, Poëtte G, Lugato D, Congedo P (2021) Goal-oriented sensitivity analysis of hyperparameters in deep learning. Preprint hal-03128298v5
55. Owen AB (2014) Sobol' indices and shapley value. SIAM/ASA J Uncertain Quant 2(1):245–251
56. Owen AB, Hoyt C (2021) Efficient estimation of the ANOVA mean dimension, with an application to neural net classification. SIAM/ASA J Uncertain Quant 9(2):708–730
57. Pearl J (2015) Generalizing experimental findings. J Causal Inf 3(2):259–266
58. Pearl J, Bareinboim E (2011) Transportability across studies: a formal approach. Tech. Rep. R-372, Department of Computer Science, University of California, Los Angeles

59. Pearl J, Bareinboim E (2014) External validity: from do-calculus to transportability across populations. Stat Sci 29(4):579–595
60. Rao CR (1985) Weighted distributions arising out of methods of ascertainment: What population does a sample represent? In: A celebration of statistics. Springer, New York, pp 543–569
61. Rasch G (1977) On specific objectivity: an attempt of formalizing the generality and validity of scientific statements. Danish Yearbook Philos 14:58–94
62. Razavi S, Jakeman A, Saltelli A, Prieur C, Iooss B, Borgonovo E, Plischke E, Piano SL, Iwanaga T, Becker W, Tarantola S, Guillaume JH, Jakeman J, Gupta H, Melillo N, Rabitti G, Chabridon V, Duan Q, Sun X, Smith S, Sheikholeslami R, Hosseini N, Asadzadeh M, Puy A, Kucherenko S, Maier HR (2021) The future of sensitivity analysis: An essential discipline for systems modeling and policy support. Env Modell Softw 137:104954
63. Ribeiro MT, Singh S, Guestrin C (2016) "why should I trust you?": Explaining the predictions of any classifier. In: Proceedings of the 22nd ACM SIGKDD international conference on knowledge discovery and data mining. ACM, New York
64. Rovelli C (2020) Helgoland. Adelphi, Milano
65. Saltelli A, Ratto M, Andres T, Campolongo F, Cariboni J, Gatelli D, Saisana M, Tarantola S (2008) Global sensitivity analysis: The primer. Wiley, Hoboken
66. Shapley LS (1953) A value for n-person games. In: Contributions to the theory of games (AM-28), vol II. Princeton University Press, Princeton, pp 307–318
67. Smith RC (2014) Uncertainty Quantification. SIAM, Philadelphia
68. Sobol I (1993) Sensitivity estimates for nonlinear mathematical models. Math Modell Comput Exp 1(4):407–414
69. Sokolić J, Giryes R, Sapiro G, Rodrigues M (2017) Generalization error of invariant classifiers. In: Proceedings of the 20th international conference on artificial intelligence and statistics, AISTATS 2017 Cited by 7
70. Stigler S (1999) Statistics on the table: The history of statistical concepts and methods. Harvard University Press, Cambridge
71. Verma S, Rubin J (2018) Fairness definitions explained. In: Proceedings of the international workshop on software fairness. ACM, Cham
72. Wang S, Jank W, Shmueli G (2008) Explaining and forecasting online auction prices and their dynamics using functional data analysis. J Bus Econ Stat 26(2):144–160
73. Wikipedia (2022) Chronology of computation of π. in wikipedia, the free encyclopedia. https://en.wikipedia.org/wiki/Chronology_of_computation_of_%CF%80. [Online; retrieved 25 February 2022]
74. Williamson B, Feng J (2020) Efficient nonparametric statistical inference on population feature importance using shapley values. In: International conference on machine learning, pp 10282–10291
75. Wolfram (2022) My pi day. Retrieved 20:18, February 27, 2022, from https://www.mypiday.com
76. Yang Q, Zhang Y, Dai W, Pan SJ (2020) Transfer Learning. Cambridge University Press, Cambridge

Chapter 2
Model Interpretability, Explainability and Trust for Manufacturing 4.0

Bianca Maria Colosimo and Fabio Centofanti

Abstract Manufacturing is currently characterized by a widespread availability of multiple streams of big data (e.g., signals, images, video-images, 3-dimensional voxel and mesh-based reconstructions of volumes and surfaces). Manufacturing 4.0 refers to the paradigm shift involving appropriate use of all this rich data environment for decision making in prognostic, monitoring, optimization and control of the manufacturing processes. The paper discusses how the new advent of Artificial Intelligence for manufacturing data mining poses new challenges on model interpretability, explainability and trust. Starting from this general overview, the paper then focuses on examples of big data mining in Additive Manufacturing. A real case study focusing on spatter modeling for process optimization is discussed, where a solution based on robust functional analysis of variance is proposed.

2.1 Manufacturing 4.0: Driving Trends for Data Mining

Industry 4.0 embraces a new generation of technologies—e.g., Internet of Things, digital twins, Artificial intelligence (AI), Robotics and Additive Manufacturing (AM)—that are paving the way to new solutions for designing, producing and supplying products and services. In this scenario, manufacturing is experiencing a paradigm shift where the digital and physical sides of processes have to be properly combined to produce a new generation of products.

With reference to data mining, new technologies are revolutionizing the way in which information is available for process optimization, monitoring and control. New sensing solutions allow a broad range of signals, image and video-images; high computational power is distributed to facilitate real-time data analysis; novel

B. M. Colosimo (✉)
Department of Mechanical Engineering, Politecnico di Milano, Milano, Italy
e-mail: biancamaria.colosimo@polimi.it

F. Centofanti
Department of Industrial Engineering, University of Naples Federico II, Naples, Italy
e-mail: fabio.centofanti@unina.it

non-contact systems are available for quality inspection of surfaces and volumes (e.g., structured light or laser scanners, X-Ray Computer Tomography); digital twins provide virtual data that can be combined with real ones to gain efficient decision making.

In this scenario, new solutions for big data modeling have to be developed, as manufacturing 4.0 involves all the V's of big data:

- Volume and Velocity are mainly ascribed to the huge amount of signals, images and video images, which can be acquired at high frequency;
- Variety is mainly due to the multiple streams of unstructured data available to describe the product and process quality: numerical data in Euclidean spaces (scalars, vectors, matrices or tensors), streaming (time based) data, network, text, and manifold data (scalars, vectors, and tensor data not on an Euclidean space);
- Veracity is eventually a crucial element to decide informative and reliable data that really represent a source of information for decision making in manufacturing.

As big data acquire strategic relevance in manufacturing, new solutions based on AI, mainly focusing on Machine Learning (ML) and Deep Learning (DL) are currently emerging as active fields of study, connecting the discipline of Statistics and Computer Science to enrich the industrial toolkit of scientists and practitioners [9].

While exploring emerging solutions and methods for statistical process monitoring and process optimization, we will explore open issues and key elements to better highlight the role of explainability and interpretability of models used in manufacturing 4.0.

2.1.1 Process Monitoring in Manufacturing 4.0

Since their origin in 1930s, methods and tools for Statistical Process Control (SPC, also known as Statistical Process Monitoring or Statistical Quality Control) are still used today. With many different variants and few exceptions (as for self-starting control charting), all the methods share a similar structure, which consists of identifying the in-control state model (in the design or training Phase, usually known as Phase 1) and then designing a control chart to detect departures from the in-control state as soon as possible. Following the technological advances of manufacturing 4.0, the effectiveness of traditional model-based SPC is undermined by many features of data available nowadays, as their high-dimensionality, their complex temporal and spatial structures that can involve a large amount of missing or corrupted data.

AI-based approaches for quality data are becoming more and more frequent, as discussed in a recent special issue of *Journal of Quality Technology* [9]. For a long time, Neural Networks (NNs) have been proposed as a substitute solution to control charts in quality monitoring [34, 40] and for prediction and classification purposes in manufacturing applications [49]. Despite the large number of papers proposing NNs for SPC, Woodall and Montgomery [45] observed that this stream of research had a limited impact in real practice. Weese et al. [43] reviewed many statistical learning methods in the field of multivariate quality control: approaches for dimension reduction, one-class classification, pattern recognition, profile monitoring, NNs and support vector machines. In their discussion, the authors showed that many of these approaches overcome some limits of traditional model-based SPC when data have highly complex structures and do not respect the model assumptions. On the other side, AI approaches pose several questions related to the difficulty in interpreting the results and diagnostic the cause behind the out-of-control alarms, and, furthermore, such methods are very limited in the sense of generalization. All these issues will be further discussed in the following.

2.1.2   Design of Experiments in Manufacturing 4.0

In manufacturing, Design of Experiments (DOE) is a critical ingredient to analyse the effect of process parameters on the response variable of interest. This is the fundamental step for process setting, process optimization, and transfer function identification when feedback control is of interest.

Active Learning (AL) is an area where ML methods have been considered to face typical challenges of DOE. AL uses a learning algorithm to select the most informative instances from the data with the goal of reducing the training data, the experimental efforts and, hence, the related costs. In analogy with DOE, AL considers the instances as experimental runs, learning instead of model fitting, and labeled instances as response values observed at each experimental condition. In this framework, explainability, interpretability and the capability to embed some pre-existing information on the process are relevant ingredients for the experimenter. Here again, black-boxes approaches are effective in relaxing assumptions for model fitting but can create relevant obstacles in interpreting the phenomena under study. As a matter of fact, parametric models can be easily structured to represent all the pre-existing knowledge on the physical principles underlying the manufacturing processes. The interpretability of the fitted model can further promote continuous improvement, as incremental knowledge of the process mechanisms can be gained after model fitting and appropriate actions can be designed to improve the process performance or better understand directions for further research. As an additional significant drawback, very large sample sizes are often needed especially for AI predictive models to achieve reasonable accuracy. This experimental effort cannot be easily available in manufacturing applications, where offline experimental data is expensive to collect and online streaming data are usually not adequate to fully gain appropriate knowledge of the process model.

As the data deluge is a real problem, classical statistics based on small samples can be ineffective and some degree of automation will be required in future processes. Manufacturing 4.0 should be hence based on a new generation of approaches which integrate AI, ML, and DL methods in non-opaque, innovative ways. Clearly, totally automatic AI algorithms to monitor, control or optimize manufacturing processes, while maybe possible to achieve in practice, are possibly dangerous. Future solutions can be envisaged where AI solutions are considered to assist decision making rather than replacing it. Additionally, even data-driven models should be assessed in practice with additional testing of the predictions using DOE and/or by monitoring the model performance. All these dimensions pave the way to a new generation of hybrid solutions where front-end knowledge with data as well as back-end testing of model performance will help to ensure that AI is used responsibly and delivers value.

2.1.3 Increasing Trust in AI Models for Manufacturing 4.0: Interpretability, Explainability and Robustness

In his seminal paper in 2001, Breiman [2] pointed out that two different cultures have been coexisting on data analysis for a long time: the first one, mainly dominated by statisticians, assuming that data are mainly generated by stochastic models that need to be estimated; the second one, mainly ruled by computer scientists, adopting algorithmic solutions to solve the problem at hand while treating the data mechanism as unknown. According to Breiman, "the statistical community has been committed to the almost exclusive use of data models" (i.e., playing in the first mainstream) leading to "irrelevant theory, questionable conclusions and keeping statisticians from working on a large range of interesting current problems." Among the attempts to combine the statistics with computer science, Statistical Learning Theory [20] and computational statistical methods [12] represent relevant examples of domains' convergence.

A similar contrast has been faced in manufacturing for a long time, but thanks to the advent of Industry 4.0, a new perspective is considered with the aim of combining statistics and computer science to develop a new generation of hybrid methodologies designed and developed with the final aim of overcoming the limits of each specific class of approaches.

On the one side, AI methods are flexible and easily adaptable to high-dimensional and complex data. Unfortunately, these advantages come at the expenses of significant drawbacks in terms of interpretability, explainability and transparency of the developed models (Fig. 2.1). The opaqueness of the AI-based solutions, in particular the black-box algorithms of "deep" learning methods, can undermine trust in safety-critical domains, reduce advances in scientific knowledge, limit the transferability of the approaches to domain shifts, and reduce the fairness of decision-making [27]. To overcome these limitations, eXplainable Artificial Intelligence (XAI) is a new field of research aiming at developing new techniques and methods to convert the so-called black-box AI algorithms to white-box algorithms, where the results achieved

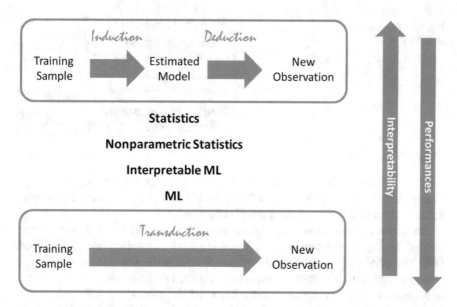

Fig. 2.1 Traditional statistics versus ML presented by Simone Vantini during the ENBIS Workshop on *Interpretability for Industry 4.0, July 12–13, 2021*. The notion of transduction, or going from specific training cases to specific test cases, was introduced in the AI/ML literature by Vapnik [41]

by the algorithms and all the parameters, variables and steps can be explained and clearly traced.

In the context of manufacturing, explainability and transferability are highly relevant. Indeed, models for manufacturing can take significant advantage of pre-existing knowledge on the underlying physics/engineering principles governing the process to be monitored, optimized or controlled. One of the main advantages of traditional statistical models, especially the parametric ones, is their identifiability and their interpretability. In manufacturing, the domain knowledge, the physics and engineering laws can be used to define the structure of the model to be fitted and to select the relevant parameters and factors to be considered. As argued by Simon [38] (Fig. 2.2), manufacturing, as many fields of engineering, fall in the broad range of sciences studying the "artificial" rather than natural phenomena. In this framework, all the pre-existing knowledge should be possibly fused with the empirical evidence gained via data analysis and this knowledge integration can be greatly simplified when interpretable solutions have to be considered.

Model explainability can further speed up process improvement, as explainable model can provide a clear understanding of the underlying phenomena to possibly suggest directions for future research.

All the aforementioned features of AI methods can be eventually summarized in the most relevant elements for data mining in Industry 4.0 applications, which is *trust*. When coming to real applications of data mining, practitioners and scientists

Fig. 2.2 The science of
artificial [38]

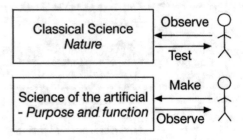

are mostly concerned about how much we can trust the obtained predictions. Clearly, explainability and interpretability contribute to model trust but a complementary feature strongly affecting trust is robustness.

Robustness is generalized by the notion of stability, which is a concept that applies throughout the entire data science life cycle, including interpretable ML [30]. The stability principle requires that each step in the life cycle is stable with respect to appropriate perturbations, such as changes in the model or data. Stability is a common sense principle, is convenient for the model interpretation, and is a prerequisite for knowledge and trustworthy interpretations. That is, one should not interpret parts of a model which are not stable to appropriate perturbations to the model and data. It is related to the notion of scientific reproducibility, which Fisher and Popper argued is a necessary condition for establishing scientific results [14, 33]. On this line Yu [48] argues that "interpretability needs stability", as the conclusions of a statistical analysis have to be robust to different levels of perturbations to be meaningful [2].

2.2 Additive Manufacturing as a Paradigmatic Example of Manufacturing 4.0

Together with Robotics and Industry of Things, AM is referred to as one of the enabling technologies of Industry 4.0, with great potential in different domains. AM refers to a broad set of emerging processes aimed at producing high-value-added products with new functional performances, as additively produced parts exhibit innovative shapes, complex features and lightweight structures that are difficult or impossible to manufacture with other processes. Design freedom and digital manufacturing are significant advantages of AM, which can be usefully considered for: (1) part customization, which is relevant for biomedical and dental applications; (2) lightweight design, which is critical to support decarbonization in the aerospace and the automotive sectors; (3) functionally-enhanced parts, which allows efficient heat exchange in oil & gas and machinery sectors.

One of the main advantage of AM is represented by its natural capability to connect the digital and the physical worlds (Fig. 2.3). In fact, products to be printed can be digitally designed (using for example, topology optimization software);

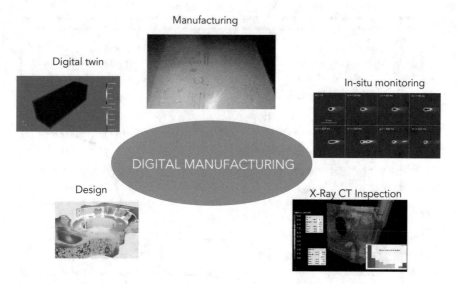

Fig. 2.3 Additive Manufacturing and the role of digital flow from the design to the final inspection

then, digital twin solutions can be used to simulate the process and in-situ signals, images and video-images can be acquired during printing to gain in-line and in-situ information on the process state. Eventually, part produced can be inspected via X-Ray Computer Tomography (X-Ray CT) to have a voxel-based reconstruction of the printed volume.

In all the aforementioned applications (e.g., aerospace, biomedical and oil & gas applications), stringent quality requirements impose defect-free and first-time-right manufacturing that cannot be easily obtained with state-of-the-art AM systems. As stated by the National Institute of Standard and Technology [25], the poor quality and high defective rates of AM processes represent one of the main barriers to the widespread adoption of this class of technologies in many industrial sectors, as

the variability in part quality due to inadequate dimensional tolerances, surface roughness, and defects, limits the metal AM broader acceptance for high-value or mission-critical applications.

Defects can be due to many different causes: improper design choices, feedstock material, problems with the equipment or unstable process conditions.

Among defects, geometrical and/or dimensional errors and volumetric errors (e.g., porosity), thermal stresses and cracks are the most common flaws. All these defects are critical, as they can significantly reduce mechanical properties of AM components (e.g., tensile and fatigue resistance).

Thanks to the layerwise nature of AM processes, big data streams can be acquired during the process itself, to keep it under control and create a digital ID card of the product based on in-situ gathered information.

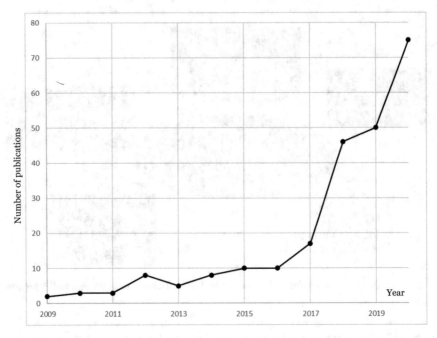

Fig. 2.4 Papers published on in-situ monitoring in Additive manufacturing with specific attention to Powder Bed Fusion processes (source Scopus)

In-situ process monitoring and process optimization can rely on this data stream [8, 19], which can be mainly classified into three data types:

- Signal data, which can be modelled as time series or profile/functional data;
- Image data, as the picture of the part printed at each layer and the surface texture observed before or after the printing step;
- Video-image data, representing the process dynamic.

As an example, Fig. 2.4 shows the number of papers published on in-situ monitoring in laser powder bed fusion, (L-PBF), the most relevant process for metal AM. It is worth observing the exponential growth of papers in this area, as an effect of the overall industrial pressure to in-situ and in-line process and product qualification and waste reduction. A detailed description of the percentages of papers focusing on different levels of in-situ monitoring is shown in Fig. 2.5, where details on the types of data involved is also highlighted. Further information on data mining in this area can be found in [19].

In the following, an example of data analysis dealing with spattering observed in video-image data is presented.

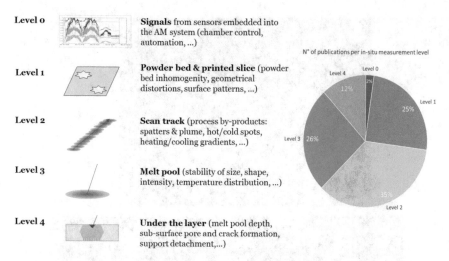

Fig. 2.5 Different levels of in-situ monitoring approaches for AM, relative percentages and types of signals

2.3   Increase Trust in Additive Manufacturing: Robust Functional Analysis of Variance in Video-Image Analysis

As previously stated, advances in data collecting technologies require new solutions to analyse a new generation of complex experimental data (i.e., images, videos, and dense point clouds) which are now common in manufacturing 4.0. Thus, a family of methods, referred to as functional data analysis (FDA) [21, 23, 35], has been introduced to model observation units as functions defined on a compact domain. A large variety of industrial applications of functional data to sensor signals and metrology data have been presented in the literature so far [3, 4, 6, 10, 29, 31, 42, 44].

This section elaborates on the work of Centofanti et al. [5] where a metal additive manufacturing process, L-PBF, is analysed by means of high speed videos acquired during the process [7, 8, 19]. L-PBF process produces metal parts with complex geometries and innovative properties by selectively melting thin layers of metal powder through a laser beam [16]. Classical statistical approaches, based on summary statistics, entail an intrinsic information loss and an arbitrary and problem dependent analysis. Centofanti et al. [5] transform the video image frames into a functional format where the quality characteristic is the *spatter intensity*, defined as the amount of spatters observed in any given region of the bi-dimensional video-frame space. In the AM application, a relevant problem consists of studying the effect of process parameters on the spatter behaviour, which is known to be a proxy of process stability and quality [1, 24].

An example of spattering is given in Fig. 2.6.

The corresponding representation of spattering considering the intensity function is shown in Fig. 2.7, which is obtained by means of a smoothing phase based

Fig. 2.6 Example of a spatter intensity function acquired during the L-PBF process

on tensor product bases of cubic splines applied to the actual amount of spatters counted in every location of the spatial domain [5].

Functional analysis of variance (FANOVA) consists of analysing the effect of parameters on the functional data, which act as response function. In general, FANOVA methods are not necessarily robust against outliers, as they rely on both the functional mean and quadratic forms, which are known to be highly sensitive to outlying observations. Centofanti et al. [5] address the problem of the FANOVA in the presence of nuisance effects associated to outliers in the sample. To deal with outliers, the *diagnostic* and the *robust* approaches are suitable choices. The former overlooks sample units identified as outliers, and is often subject to the analyst's subjective decision. The robust approach, on the other hand, automatically generates parameter estimators, as well as related tests and confidence intervals, that reduce the impact of outliers on final results. As discussed in Sect. 2.1.3, robustness is needed for the model interpretation and is a prerequisite for trustworthy interpretations.

In the very last years, several works have explored robust estimation for functional data [11, 15]. Sinova et al. [39] present the M-estimators for functional data, which is applied by Kalogridis and Van Aelst [22] to the functional linear model. However, to the best of the authors' knowledge, Centofanti et al. [5] is the

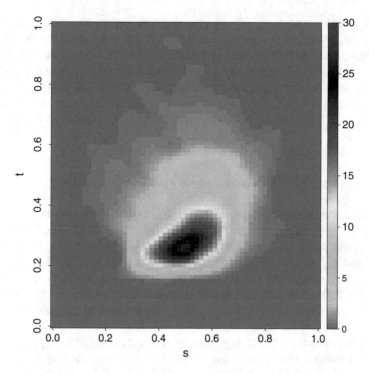

Fig. 2.7 Example of a spatter intensity function acquired during the L-PBF process

first robust ANOVA method for functional data. They propose a robust functional ANOVA method (RoFANOVA) that is able to test, in a nonparametric fashion, the differences among group functional means. The RoFANOVA method is based on a functional generalization of the test statistic proposed by Schrader and Mc Kean [37] included in a permutational framework [17, 32]. The RoFANOVA method is implemented in the R package rofanova, openly available on CRAN.

2.3.1 The RoFANOVA Approach

The RoFANOVA approach elaborates on the functional M-estimator proposed by Sinova et al. [39], defined as

$$\hat{\mu}_s = \underset{y \in L^2(\mathcal{T})}{\mathrm{argmin}} \sum_{i=1}^{n} \rho\left(\|X_i - y\|\right), \tag{2.1}$$

where $\rho : \mathbb{R}^+ \rightarrow \mathbb{R}$ is the *loss function*, which is continuous, non-decreasing and satisfies $\rho(0) = 0$. Starting from this, Centofanti et al. [5] propose a scale equivariant version of M-estimator that relies on the FuNMAD estimator of dispersion, which is the functional extension of the normalized median absolute deviation [28]. The scale equivariant functional M-estimator has not a closed-form solution and a numerical solution can be obtained through the standard iteratively re-weighted least-squares algorithm proposed by Sinova et al. [39]. The loss function ρ in Equation (2.1) defines the properties of the resulting estimator and several choices are available as the Huber, Tukey, Hampel, and optimal families of loss functions [5].

The two-way functional ANOVA model with interaction is

$$X_{ijk}(t) = m(t) + f_i(t) + g_j(t) + h_{ij}(t) + \varepsilon_{ijk}(t) \quad t \in \mathcal{T}, \tag{2.2}$$

where m is the functional grand mean, which reflects the process's overall form, f_i and g_j are the functional main effects and h_{ij} is the interaction term. The functions X_{ijk}, are the realizations of a functional quality characteristic X, defined on $\mathcal{T}$, at level i and j, $i = 1, \ldots, I$ $j = 1, \ldots, J$, of the two factors A and B, that affect X, and ε_{ijk} is a functional error term. The significance of the coefficients in the model in Equation (2.2) is tested through the standard null and alternative hypotheses, which test the significance of the factors A and B, and the interaction term, respectively [35].

Each test is carried out through a nonparametric permutational approach based on the functional extension of the robust F-statistic proposed by Schrader and Mc Kean [37], which is a robust version of the classical F-test statistic used in the multivariate setting. The test statistic is the difference between the standardized residual sum of dispersions under the reduced and the full model and measures the discrepancy between residuals of the model under the null and the alternative hypotheses. The distribution of the considered statistic under the null hypothesis is estimated through the permutational approach of Manly [26] that consists of permuting the data without any constraints.

2.3.2 An Additive Manufacturing Application

In L-PBF, the spatters can be expelled either by the melt pool, i.e., the melting region, or by the regions nearby the melt pool [1, 24, 47]. Spatter analysis in the L-PBF process has got a lot of attention in recent years since it provides useful information about the current condition of the process and the quality of the manufactured item. The influence of controllable process parameters and other operational circumstances on spatter behaviour may be used to gain better understanding of the physical phenomena at hand. Such information may be utilized to fine tune the process conditions and improve product quality and mechanical performance, or to develop in-line and real-time process monitoring systems. High-speed cameras, deployed inside or outside the viewports of the L-PBF machine, may detect hot spatters expelled as a result of the laser-material interaction. The major research on L-PBF spatter analysis and monitoring uses video image processing methods to provide synthetic indices to capture key characteristics of spatter behaviour, such as the number of ejected spatters in each video frame, their size, speed, and so on [13, 18]. Instead of interpreting synthetic descriptors of spatter ejections as univariate or multivariate variables, the spatter behaviour is fully characterized here by the so-called spatter intensity function, denoted by Y_{ijk} ($i = 1 \ldots, 6$, $j = 1, \ldots 6$ and $k = 1, \ldots, n_{ij}$). In the application, the production of six specimens via L-PBF of maraging steel powder is considered. Specimens are produced by varying the energy density provided by the laser to the material for six production layers. For each layer, the laser moves along a predefined path consisting of parallel scan lines, whose orientation changes layer by layer. The spatter intensity functions are modelled according to Equation (2.2), where f_i is the effect of the energy density level, g_i is the layer effect, and h_{ij} is the interaction term. The analysis aims to asses the significance of the energy density level, the layer, and their interaction term. The cross-sections of the spatter intensity functions are represented in Fig. 2.8 at different energy density levels and in different layers.

The L-PBF process has complex dynamics with several transient and local occurrences that may alter the inherent variability of the measured quantities and lead to outlying patterns. Determining whether or not an experimental point is an outlier and the diagnosis of root causes may result practically infeasible in the vast majority of innovative processes as the one at hand. In these cases, the RoFANOVA test can be suitably applied. Several loss functions are considered and tuned to achieve the 95% asymptotic efficiency at the normal model. All tests agree that the energy density and the layer effects are significant. Moreover, the energy density is increasing with the intensity of the spatter, which is consistent with the fact that a higher energy density results in a larger, hotter melt pool with stronger convective and rebound processes [1, 36, 46]. Moreover, different layers show different increasing patterns. Indeed, in layers 1, 2, and 6, the spatter intensity is increasing with respect to the energy density. The laser scan directions of these three levels are very similar. A scan direction parallel to the gas flow increases the number of powder bed particles that are pushed down the laser path and heated up by the hot metal vapour emission as well as those expelled as hot spatters. In this condition, the larger the energy density, the stronger the convective motions that entrap the powder particles in the hot vapour emission, and thus the spatter intensity. In layers 3, 4, and 5, the energy density had a different effect on the spatter intensity. In these layers, the laser scan direction was in fact almost perpendicular to the shielding gas flow direction. Particles are drawn away from the scan path under these conditions, so reducing the number of powder particles ejected as hot spatters and, as a result, the overall spatter intensity.

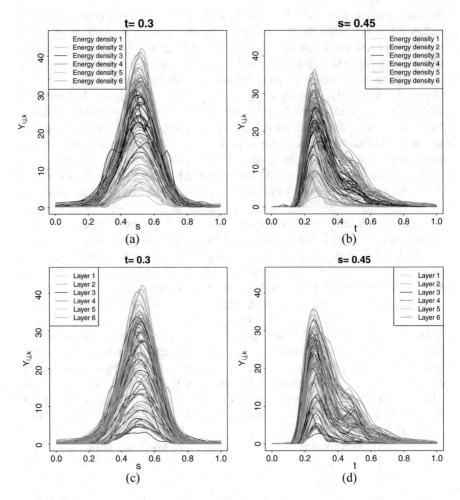

Fig. 2.8 Cross-sections of the spatter intensity functions $Y_{i,j,k}$ for $t = 0.3$, (**a**) and (**c**), and $s = 0.45$, (**b**) and (**d**), in the real-case study, for different energy density levels ((**a**) and (**b**)) and different layers ((**c**) and (**d**))

Acknowledgments Authors are happy to acknowledge Marco Luigi Grasso, Alessandra Menafoglio, Biagio Palumbo and Simone Vantini for their contribution.

References

1. Bidare P, Bitharas I, Ward R, Attallah M, Moore AJ (2018) Fluid and particle dynamics in laser powder bed fusion. Acta Mater 142:107–120
2. Breiman L (2001) Statistical modeling: The two cultures (with comments and a rejoinder by the author). Stat Sci 16(3):199–231

3. Capezza C, Centofanti F, Lepore A, Palumbo B (2021) Functional clustering methods for resistance spot welding process data in the automotive industry. Appl Stoch Models Business Ind 37(5):908–925
4. Capezza C, Centofanti F, Lepore A, Menafoglio A, Palumbo B, Vantini S (2022) Functional regression control chart for monitoring ship CO_2 emissions. Qual Reliab Eng Int 38(3):1519–1537
5. Centofanti F, Colosimo BM, Grasso ML, Menafoglio A, Palumbo B, Vantini S (2021) Robust functional ANOVA with application to additive manufacturing. Preprint. arXiv:211210643
6. Centofanti F, Lepore A, Menafoglio A, Palumbo B, Vantini S (2021) Functional regression control chart. Technometrics 63(3):281–294
7. Colosimo BM, Grasso M (2020) On-machine measurement, monitoring and control. CRC Press
8. Colosimo BM, Huang Q, Dasgupta T, Tsung F (2018) Opportunities and challenges of quality engineering for additive manufacturing. J Qual Technol 50(3):233–252
9. Colosimo BM, del Castillo E, Jones-Farmer LA, Paynabar K (2021a) Artificial intelligence and statistics for quality technology: an introduction to the special issue. J Qual Technol 53(5):443–453
10. Colosimo BM, Grasso M., Garghetti F, Rossi B (2021) Complex geometries in additive manufacturing: A new solution for lattice structure modeling and monitoring. J Qual Technol 1–23
11. Cuesta-Albertos JA, Fraiman R (2006) Impartial trimmed means for functional data. DIMACS Series in Discrete Mathematics and Theoretical Computer Science 72:121
12. Efron B, Hastie T (2016) Computer age statistical inference, vol 5. Cambridge University Press
13. Everton SK, Hirsch M, Stravroulakis P, Leach RK, Clare AT (2016) Review of in-situ process monitoring and in-situ metrology for metal additive manufacturing. Mater Des 95:431–445
14. Fisher RA (1936) Design of experiments. Br Med J 1(3923):554
15. Fraiman R, Muniz G (2001) Trimmed means for functional data. Test 10(2):419–440
16. Gibson I, Rosen D, Stucker B, Khorasani M (2014) Additive manufacturing technologies, vol 17. Springer, New York
17. Good P (2013) Permutation tests: a practical guide to resampling methods for testing hypotheses. Springer Science & Business Media
18. Grasso M, Colosimo BM, Tsung F (2017) A phase I multi-modelling approach for profile monitoring of signal data. Int J Prod Res 55(15):4354–4377
19. Grasso M, Remani A, Dickins A, Colosimo B, Leach R (2021) In-situ measurement and monitoring methods for metal powder bed fusion: An updated review. Meas Sci Technol 32(11)
20. Hastie T, Tibshirani R, Friedman JH, Friedman JH (2009) The elements of statistical learning: data mining, inference, and prediction, vol 2. Springer, New York
21. Horváth L, Kokoszka P (2012) Inference for functional data with applications. Springer Science & Business Media
22. Kalogridis I, Van Aelst S (2019) Robust functional regression based on principal components. J Multivariate Anal 173:393–415
23. Kokoszka P, Reimherr M (2017) Introduction to functional data analysis. CRC Press
24. Ly S, Rubenchik AM, Khairallah SA, Guss G, Matthews MJ (2017) Metal vapor micro-jet controls material redistribution in laser powder bed fusion additive manufacturing. Sci Rep 7(1):1–12
25. Mani M, Lane B, Donmez M, Feng S, Moylan S, Fesperman R (2015) Measurement science needs for real-time control of additive manufacturing powder bed fusion processes. NIST Interagency/Internal Report (NISTIR), National Institute of Standards and Technology, Gaithersburg, MD
26. Manly BF (2006) Randomization, bootstrap and Monte Carlo methods in biology, vol 70. CRC Press
27. Marcinkevičs R, Vogt JE (2020) Interpretability and explainability: A machine learning zoo mini-tour. Preprint. arXiv:201201805

28. Maronna RA, Martin RD, Yohai VJ, Salibián-Barrera M (2019) Robust statistics: theory and methods (with R). John Wiley & Sons
29. Menafoglio A, Grasso M, Secchi P, Colosimo B (2018) Profile monitoring of probability density functions via simplicial functional PCA with application to image data. Technometrics 60(4):497–510
30. Murdoch WJ, Singh C, Kumbier K, Abbasi-Asl R, Yu B (2019) Definitions, methods, and applications in interpretable machine learning. Proc Natl Acad Sci 116(44):22071–22080
31. Noorossana R, Saghaei A, Amiri A (2011) Statistical analysis of profile monitoring, vol 865. John Wiley & Sons
32. Pesarin F, Salmaso L (2010) Permutation tests for complex data: theory, applications and software. John Wiley & Sons
33. Popper K (2005) The logic of scientific discovery. Routledge
34. Psarakis S (2011) The use of neural networks in statistical process control charts. Qual Reliab Eng Int 27(5):641–650
35. Ramsay JO, Silverman BW (2005) Functional data analysis. Springer, New York
36. Repossini G, Laguzza V, Grasso M, Colosimo BM (2017) On the use of spatter signature for in-situ monitoring of laser powder bed fusion. Additive Manuf 16:35–48
37. Schrader RM, Mc Kean JW (1977) Robust analysis of variance. Commun Stat Theory Methods 6(9):879–894
38. Simon HA (1996) The sciences of the artificial. MIT Press
39. Sinova B, Gonzalez-Rodriguez G, Van Aelst S, et al (2018) M-estimators of location for functional data. Bernoulli 24(3):2328–2357
40. Thomson A (1988) Real-time artificial intelligence for process monitoring and control. IFAC Proc Vol 21(13):67–72
41. Vapnik VN (1998) Statistical learning theory. John Wiley & Sons
42. Wang K, Tsung F (2005) Using profile monitoring techniques for a data-rich environment with huge sample size. Qual Reliab Eng Int 21(7):677–688
43. Weese M, Martinez W, Megahed FM, Jones-Farmer LA (2016) Statistical learning methods applied to process monitoring: An overview and perspective. J Qual Technol 48(1):4–24
44. Wells L, Megahed F, Niziolek C, Camelio J, Woodall W (2013) Statistical process monitoring approach for high-density point clouds. J Intell Manuf 24(6):1267–1279
45. Woodall WH, Montgomery DC (2014) Some current directions in the theory and application of statistical process monitoring. J Qual Technol 46(1):78–94
46. Yang L, Lo L, Ding S, Özel T (2020) Monitoring and detection of meltpool and spatter regions in laser powder bed fusion of super alloy Inconel 625. Prog Additive Manuf 5(4):367–378
47. Young ZA, Guo Q, Parab ND, Zhao C, Qu M, Escano LI, Fezzaa K, Everhart W, Sun T, Chen L (2020) Types of spatter and their features and formation mechanisms in laser powder bed fusion additive manufacturing process. Additive Manuf 36:101438
48. Yu B (2013) Stability. Bernoulli 19(4):1484–1500
49. Zhang HC, Huang S (1995) Applications of neural networks in manufacturing: a state-of-the-art survey. Int J Prod Res 33(3):705–728

Chapter 3
Interpretability via Random Forests

Clément Bénard, Sébastien Da Veiga, and Erwan Scornet

Abstract Although there is no consensus on a precise definition of interpretability, it is possible to identify several requirements: "simplicity, stability, and accuracy", rarely all satisfied by existing interpretable methods. The structure and stability of random forests make them good candidates to improve the performance of interpretable algorithms. The first part of this chapter focuses on rule learning models, which are simple and highly predictive algorithms, but very often unstable with respect to small data perturbations. A new algorithm called SIRUS, designed as the extraction of a compact rule ensemble from a random forest, considerably improves stability over state-of-the-art competitors, while preserving simplicity and accuracy. The second part of this chapter is dedicated to post-hoc methods, in particular variable importance measures for random forests. An asymptotic analysis of Breiman's MDA (Mean Decrease Accuracy) shows that this measure is strongly biased using a sensitivity analysis perspective. The Sobol-MDA algorithm is introduced to fix the MDA flaws, replacing permutations by projections. An extension to Shapley effects, an efficient importance measure when input variables are dependent, is then proposed with the SHAFF algorithm.

C. Bénard (✉)
Safran Tech, Digital Sciences & Technologies, Magny-Les-Hameaux, France

Sorbonne Université, CNRS, LPSM, Paris, France
e-mail: clement.benard@safrangroup.com

S. Da Veiga
Safran Tech, Digital Sciences & Technologies, Magny-Les-Hameaux, France
e-mail: sebastien.da-veiga@safrangroup.com

E. Scornet
École Polytechnique, Institut Polytechnique de Paris, CMAP, Palaiseau, France
e-mail: erwan.scornet@polytechnique.edu

© The Author(s), under exclusive license to Springer Nature Switzerland AG 2022
A. Lepore et al. (eds.), *Interpretability for Industry 4.0 : Statistical and Machine Learning Approaches*, https://doi.org/10.1007/978-3-031-12402-0_3

3.1 Introduction

State-of-the-art learning procedures, such as tree ensembles or neural networks, are acknowledged for their outstanding predictive performance. However, the quality of their predictions results from complex mechanisms: each prediction is the fruit of a large number of elementary operations. Because of this apparent complexity, most state-of-the-art learning algorithms are considered as black boxes. This absence of interpretability is a major limitation for many applications involving critical decisions, such as healthcare, criminal justice, or industrial process optimization. Healthcare is an interesting domain to illustrate the essential aspect of interpretability [60]. Indeed, a doctor must understand the rationale behind the algorithm recommendation of a given treatment before he can legitimately prescribe it, for both practical efficiency and ethical reasons. Interpretability allows us to validate and trust the treatment, which is necessary to apply it in practice. Overall, interpretability is a property which is both essential and difficult to fulfill.

As stated by Rüping [80], Lipton [61], Doshi-Velez and Kim [31], or Murdoch et al. [71], to date, there is no clear consensus in statistics and machine learning communities about a precise definition of interpretability. Interpretability consists in multiple broad concepts, involves several types of methods, and its very definition heavily depends on the application domain and the targeted audience. Therefore, it is very difficult to provide a generic definition of interpretability valid for all application areas. In the healthcare example, we seek to interpret a given prediction prior to its application. For other types of problems, one may be interested in understanding how the inputs of a system are related to the system output, in order to change the input settings to influence the output values. A typical case is the optimization of industrial processes. Indeed, manufacturing lines often involve complex physical and chemical phenomena to transform materials. The control and efficiency of these industrial processes are of critical importance for the final quality of the production. A manufacturing line can be summarized as a sequence of transforming operations, controlled by a high number of input variables which define the industrial process: temperatures, pressures, times, weights... At the end of the line, quality tests are performed to check that each produced entity has the desired quality. The engineer objective is to find the production conditions generating defects, in order to avoid them thanks to a better setting of the input variables, and thus to improve the efficiency of the production process. Because of the complexity of these processes, they can involve hundreds of variables. Therefore, an approach based on algorithms fed with the data collected along the manufacturing line, has a critical impact in practice. Indeed, the retrieved information enables to infer a link between the manufacturing conditions and the resulting quality at the end of the line, using learning algorithms, and then, to ultimately improve the process efficiency by a better setting of the input variables.

Despite the lack of definition of interpretability, we argue that it is possible to define minimum requirements for interpretability through the triptych "simplicity, stability, and accuracy" [7], following the recent framework of Yu and Kumbier

[102]. More precisely, the links between the inputs and the output must be simple to comprehend their relation. The general concept of simplicity results from the notion of interpretability [e.g., 37, 59–61, 71, 77, 80] and essentially refers to the number of operations performed in the prediction mechanism of an interpretable model, or the complexity of the output of a post-processing method. Murdoch et al. [71] define several properties to discuss simplicity more precisely in the case of interpretable models: sparsity, simulatability, and modularity. A sparse model is based only on a small fraction of the input variables. A model is simulatable if a human can reproduce the entire prediction mechanism by hand. This is a strong restriction on the model shape, and simulatable models achieve a good predictivity only if the relation in the data is quite simple, which is not the case for image recognition for example. A model is modular if a portion of it can be interpreted independently. Modularity is a weaker constraint on the model form than sparsity or simulatability, but modular models are not as easy to understand. Secondly, in the statistical learning theory, stability of supervised algorithms refers to the stability of predictions [94]. In particular, Rogers and Wagner [79], Devroye and Wagner [30], and Bousquet and Elisseeff [10] outline that stability and predictive accuracy are closely connected. When addressing interpretability issues, stability has a broader meaning according to Yu [101] and is, in fact, another fundamental requirement for interpretability: in order to be meaningful, results from statistical analyses must be robust to small data perturbations. Indeed, a specific analysis is likely to be run multiple times, eventually adding a small new batch of data, and an interpretable algorithm should be insensitive to such minor modifications. Otherwise, unstable methods provide us with a partial and arbitrary analysis of the underlying phenomena, and arouse distrust of the domain experts. Lastly, if the predictive accuracy of an interpretable model is significantly lower than the one of a state-of-the-art black-box algorithm, or if a post-processing method gives bias results, we clearly miss strong patterns in the data and will obtain misleading conclusions, as explained by Breiman [14]. For instance, the trivial estimator that always outputs the empirical mean of the observations is simple, stable, but does not bring in most cases any useful information. Consequently, we add a good accuracy as an essential requirement for interpretability.

There are two main approaches to obtain interpretable algorithms: post-treat a black-box model to understand its prediction mechanism, or initially constrain the algorithm to have a simple structure relating inputs to the output in a clear fashion [48, 71]. Firstly, we present intrinsically interpretable rule models in Sect. 3.2. Secondly, Sect. 3.3 is dedicated to post-hoc interpretable methods via variable importance measures. Both interpretable models and post-hoc methods have flaws in regards of the interpretability requirements: simplicity, stability, and accuracy. Indeed, interpretable models are often unstable because of the strong constraints on the model shape, while post-hoc methods are often inaccurate when input variables are dependent. In both approaches, we will see that the structure and stability of random forests [13] can be used to improve upon existing algorithms.

Throughout this chapter, we use the standard supervised learning framework, with a real input vector $\mathbf{X} = (X^{(1)}, X^{(2)}, \ldots, X^{(p)}) \in \mathbb{R}^p$ of dimension p, and an output Y, which is real and continuous in the regression case, and categorical for classification problems. Additionally, we have access to a dataset $\mathcal{D}_n =$ $\{(\mathbf{X}_1, Y_1), \ldots, (\mathbf{X}_n, Y_n)\}$ of n independent pairs of random variables, distributed as $(\mathbf{X}, Y)$, used to train the learning algorithm of interest. For the sake of clarity, we mainly restrict the mathematical formulation of the following methods to the case of numerical variables, but most of the presented algorithms can also handle categorical inputs.

3.2 Interpretable Rule-Based Models

A first approach to obtain interpretable machine learning algorithms is to choose a model belonging to a class of functions with a simple structure that makes it intrin-sically interpretable in the first place. In such approaches, the *simplicity* condition of the previous triptych (simplicity, stability, accuracy) is trivially satisfied, and does not prevent to reach a good accuracy for a wide range of applications. On the other hand, stability is usually the main flaw of interpretable models because of their simple structure. This phenomenon is characterized as the "Rashomon effect" by Breiman [14]: within a class of simple models, there are many equally good models, and one is arbitrarily picked by the algorithm heuristic. When data is perturbed, the returned model changes, which explains the unstable behavior of interpretable models. Overall, there are mainly four types of intrinsically interpretable algorithms: parametric models, additive models, decision trees, and rule models. In some applications, systems are too complex to use parametric models, and additive models are more difficult to interpret than rule-based models. In this section, we focus on rule models, and also present decision trees, as these two algorithm types are strongly connected. We start by reviewing the existing rule models present in the literature in Sect. 3.2.1. Then, we introduce in Sect. 3.2.2 a recent rule algorithm, SIRUS (Stable and Interpretable RUle Set) [5, 7], and demonstrate that SIRUS can address classification and regression problems efficiently, while producing compact and stable list of rules.

3.2.1 Literature Review

3.2.1.1 Definitions and Origins of Rule Models

Origins Rule learning can be traced back to 1969 with Michalski's AQ system [70], and was a very active research area in the 1980s and 1990s. A rule learning algorithm takes the form of a collection of rules. Each rule is an if-then statement: if a hard condition on the input variables is satisfied, it implies a given value for the output. A rule can also be seen as a hypercube in the input parameter space with a constant output (see Fig. 3.1).

Fig. 3.1 Example of a decision rule, based on variables $X^{(1)}$ and $X^{(3)}$

$$\text{If } \begin{cases} X^{(1)} < 1.12 \\ \text{and } X^{(3)} \geq 0.7 \end{cases} \text{ then } \hat{Y} = 0.18 \,.$$

Originally, rule learning algorithms were mostly limited to classification problems, and were extended to regression in the 1990s. At the end of this decade, the research activity in rule learning declined, and the machine learning community focused more on improving black-box models. In the past fifteen years, there has been a renewed interest in rule learning models and their strong interpretability properties. There are three ways of combining a collection of rules to form a rule model: disjunctive normal form (DNF), decision list, and weighted rule ensemble. Firstly, DNF only deals with binary classification problems, and is based on the "separate-and-conquer" principle. One class is selected, and each rule covers a portion of the observations of the selected class with no overlap between rules. If a new data point satisfies a rule, the associated class is predicted, otherwise the default class is returned. Secondly, decision lists have a hierarchical structure and rules are ordered. Thus, a prediction is made by the first rule of the list satisfied by a new query point. Thirdly, a weighted rule ensemble assigns a weight to each rule of the collection, and can handle regression problems. A prediction is made by adding the weight of all rules satisfied by a a new query point. A high number of variants of rule learning algorithms were developed, and an exhaustive survey of DNF and decision lists based on the separate-and-conquer principle was conducted by [43]. Main DNF, decision list, and weigthed rule ensemble algorithms are presented below.

Disjunctive Normal Forms Key DNF algorithms are AQ system from Michalski [70], IREP from Fürnkranz and Widmer [44] and RIPPER from Cohen [23]. IREP builds rules sequentially following the separate-and-conquer principle, i.e., covered data points are removed from the data before learning the next rule. Each rule is fitted with a greedy heuristic: elementary constraints are added one by one to maximize a given loss function on a training set. Then, the rule is pruned back to maximize its accuracy on a testing set by removing constraints one by one. RIPPER improves IREP by adding a post-treatment step to optimize the rule list: the final IREP rule list is perturbed iteratively and the best list is picked step by step.

Decision Lists Decision lists were initially developed by Rivest [78] as an extension of DNF, and are more expressive than DNF since they are a more flexible class of learning algorithms. The learning procedure is also greedy and based on the separate-and-conquer principle. Each rule is built to classify all training examples perfectly. CN2 is a decision list developed by Clark and Niblett [22], where rules are selected to maximize predictive accuracy, based on the mechanisms of ID3 and AQ algorithms. Figure 3.2 provides an example of a decision list using a recent algorithm applied to the Titanic dataset.

Weighted Rule Ensemble Naturally, bagging [11] and boosting [38] were applied to rule learning and led to many improvements in the late 1990s. Following

if	$\begin{cases} \text{male} \\ \textbf{and } \text{adult} \end{cases}$	**then**	survival probability = 0.21 (0.19 - 0.23)
else if	3rd class	**then**	survival probability = 0.44 (0.38 - 0.51)
else if	1rd class	**then**	survival probability = 0.96 (0.92 - 0.99)
else	survival probability = 0.88 (0.82 - 0.94) .		

Fig. 3.2 Decision List example for the Titanic dataset (confidence intervals in brackets) [60, BRL]

the separate-and-conquer principle, covered data points are removed step by step in a DNF construction. Instead of removing covered data points, their weights were alleviated as boosting suggests, generating weighted rule lists with improved predictive performances. Cohen and Singer [24] applied boosting to IREP to developed SLIPPER, a weighted rule list which overperforms RIPPER. Weiss and Indurkhya [97] used boosting to develop LRI, a DNF which handles multiple class classification. LRI produces a DNF with an equal number of rules for each class. Unlike most of the previous rule learning algorithms, LRI does not use pruning but limits the complexity of each rule in the learning process, with a direct reference to the work of Friedman on boosting [39] where tree depth is limited.

The connection between decision trees and rule algorithms started in 1987 with Quinlan [75], who proposed to extract rules from a decision tree, and was extensively revisited recently. In order to explain these methods, let us first introduce decision trees.

3.2.1.2 Decision Trees

Decision trees are supervised learning algorithms, which follow the structure of a binary tree to partition the input space. Because of this specific structure, the tree predictive process is especially easy to compute by hand, and trees are therefore good candidates when interpretability is required. Trees were made popular by Breiman et al. [16] with CART for both regression and classification problems, and Quinlan [74] with ID3 for classification. These two algorithms differ in terms of splitting and stopping criteria, as ID3 is based on entropy. Here, we focus on CART, and first present the regression case.

The main principle of decision trees is to recursively partition the input space with splits of the form $X^{(1)} < z$, where z is a real threshold—see Fig. 3.3, using the training data $\mathscr{D}_n$. The observations of a given tree node are separated in two children nodes with a split of the above form, and this is recursively repeated down the tree. The tree growing is stopped such that all terminal leaves contain at least a number `min_node_size` of observations, an hyperparameter of the algorithm. A fully grown tree is likely to strongly overfit the data, and a pruning procedure [35] is usually applied after the tree growing to remove non-significant splits, identified

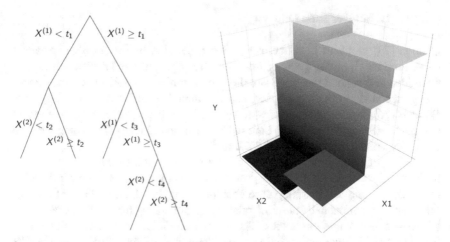

Fig. 3.3 Example of a decision tree and the associated function for $p = 2$, inspired from Friedman et al. [40]

by cross-validation. To compute a prediction for a new query point $\mathbf{x} \in \mathbb{R}$, we first drop $\mathbf{x}$ down the tree until it reaches a terminal leaf. Then, the tree estimate $m_n(\mathbf{x})$ is the average of the Y_i for the training observations belonging to the same terminal leaf, where $A_n(\mathbf{x})$ is the terminal leaf of the tree in which $\mathbf{x}$ belongs. Overall, a decision tree forms a piecewise constant estimate, as shown in Fig. 3.3. CART natively handles categorical variables with splits of the form $X^{(1)} \in \{a, b\}$, if a and b are categorical values that $X^{(1)}$ can take. The tree is constructed node by node in a greedy fashion using $\mathscr{D}_n$. Each split selects a variable and a threshold to maximize the CART splitting criterion, which measures the decrease of output variance between the parent node and the children nodes. Besides, CART also handles classification problems, where the output Y takes categorical values. In this case, the prediction process and the splitting criterion are adapted. The prediction for a new query point $\mathbf{x}$ is the majority class among the training points which fall in the same terminal leaf as $\mathbf{x}$. Besides, Gini index is used as CART splitting criterion.

Breiman [14, page 206] observes that decision trees are unstable: "if the training set is perturbed only slightly, say by removing a random 2–3% of the data, I can get a tree quite different from the original". He claims that stabilizing the structure of interpretable models is impossible because of what he called the "Rashomon effect": within a class of models, there is a high number of equally good ones. Therefore one cannot expect uniqueness: CART is a greedy heuristic, generating an arbitrary good model, unstable by construction.

3.2.1.3 Tree-Based Rule Learning

Quinlan [75] proposes to extract rules from a decision tree to form an ensemble model, and thus building a connection between rule learning and decision trees. The main idea is that each tree node is defined as a conjunction of splits which forms a hyperrectangle in the input space, and can therefore define an elementary rule. Quinlan [76] developed C4.5 rules, a weighted rule list. Rules are extracted from a decision tree and then pruned, both individually and globally. The computational efficiency was later improved with C5.0 in 1997.

The resurgence of rule learning was essentially initiated by Friedman et al. [42] who developed RuleFit: the post-processing of a tree ensemble method—importance sampling learning ensemble [41]—by the Lasso [93] enables the selection of a quite small subset of rules while preserving the predictive accuracy of state-of-the-art tree ensembles. Therefore, RuleFit shows that the structure of tree-based black-box models could be considerably simplified while preserving the same level of predictivity. More precisely, while a random forest typically runs ten thousands operations to compute a prediction, RuleFit runs about fifty operations. Thus, RuleFit was claimed to be an interpretable technique. However, there are two strong limitations to the interpretability of RuleFit. Firstly, there is a high redundancy in the output list of rules: some variables and variable interactions are involved in many rules, and some rules are very similar or highly correlated—see Fig. 3.4.

Secondly, the algorithm is highly unstable to small data perturbations: even running RuleFit on exactly the same dataset leads to quite different lists of rules.

if	$\begin{cases} \text{DIS} < 1.40 \\ \textbf{and } \text{PTRATIO} > 17.9 \\ \textbf{and } \text{LSTAT} < 10.5 \end{cases}$	**then**	$\hat{Y} = 10.1$	**else** $\hat{Y} = 0$
if	$\begin{cases} \text{RM} > 6.62 \\ \textbf{and } \text{NOX} < 0.67 \end{cases}$	**then**	$\hat{Y} = 2.26$	**else** $\hat{Y} = 0$
if	$\begin{cases} \text{RM} < 7.45 \\ \textbf{and } \text{DIS} > 1.37 \\ \textbf{and } \text{TAX} > 219.0 \end{cases}$	**then**	$\hat{Y} = -2.27$	**else** $\hat{Y} = 0$
if	$\begin{cases} \text{DIS} > 1.30 \\ \textbf{and } \text{PTRATIO} > 19.4 \end{cases}$	**then**	$\hat{Y} = -1.40$	**else** $\hat{Y} = 0$
if	$\begin{cases} \text{RM} > 7.44 \\ \textbf{and } \text{PTRATIO} < 17.9 \end{cases}$	**then**	$\hat{Y} = 2.58$	**else** $\hat{Y} = 0$
if	$\begin{cases} \text{RM} > 6.64 \\ \textbf{and } \text{NOX} < 0.67 \end{cases}$	**then**	$\hat{Y} = 1.30$	**else** $\hat{Y} = 0$
if	$\begin{cases} \text{RM} > 7.45 \\ \textbf{and } \text{PTRATIO} < 19.7 \end{cases}$	**then**	$\hat{Y} = 2.15$	**else** $\hat{Y} = 0$

Fig. 3.4 Example of RuleFit on the Boston Housing data [40], where the output Y is the median house value in a given neighborhood

Indeed, ISLE is a randomized procedure and boosting propagates perturbations in the rule generation. Furthermore, instability is amplified by the Lasso in this context of high correlation. Therefore, RuleFit violates the stability principle of interpretability, and may only be efficient for local interpretations: it is possible to retrieve the small set of rules satisfied by a specific new query point, and to add their weights to generate the prediction. Many algorithms were derived from RuleFit, but none of them directly tackle this issue of structure stability. A first type of improvement is the replacement of Lasso by other regression techniques, for example the horseshoe prior [72], well known to give aggressive shrinkage to noise predictors. It removes rules with a high number of splits or with small support, increasing the simplicity of the fitted model and the predictive accuracy. Another extension is proposed by Meinshausen [67] with Node harvest. He replaced ISLE by a random forest and applied a constrained quadratic program to the extracted rules to build a sparse rule model. Node harvest is also unstable, and outperforms the predictivity of RuleFit only in high dimension or on noisy data. Finally, we also mention the approach of Liu et al. [62], which design CRF, an algorithm combining rule extraction and feature elimination. Rules are extracted from a random forest and selected via a linear program. Selected features are used to build the following forest and these two steps are repeated until convergence.

3.2.1.4 Modern Rule Learning

Besides tree-based rule learning, traditional greedy heuristics were also quite recently extended to improve rule algorithm efficiency. Indeed, Dembczyński et al. [29] developed ENDER, a general statistical learning framework to build boosted weighted rule lists. MLRules [28] is an instance of ENDER which builds a rule ensemble by greedily minimizing the log likelihood, and each rule is built adding elementary constraints one by one. ENDER has a significantly better predictive accuracy than SLIPPER and RuleFit, but the difference with LRI was not significant Dembczyński et al. [29]. Overall, LRI, SLIPPER, and RuleFit were not significantly better to any other. In most rule learning heuristics, elementary constraints are greedily combined to form each rule. Malioutov and Varshney [66] proposed a different approach by using a linear program (LP) to learn each rule of a DNF. The LP objective function aims to minimize the number of constraints in the rule but not to maximize the rule predictivity. It includes a tuning parameter which is fixed in practice because the algorithmic complexity of a LP does not allow a fine tuning. However, predictive accuracy is no better than CART. Later, Su et al. [91] refine the LP formulation and the resulting algorithm reaches a better predictive accuracy. Finally, Letham et al. [60] developed Bayesian Rule List, a Bayesian approach of decision lists. The model is generative and the prior encourages strong sparsity leading to very simple rule models—see the example for the titanic dataset in Fig. 3.2. Note that Yang et al. [100] developed a scalable implementation of BRL.

3.2.2   SIRUS: Stable and Interpretable RUle Set

Now, we introduce a stable rule algorithm, SIRUS (Stable and Interpretable RUle Set) [5, 7], and, as a consequence, we show that rule methods can solve classification and regression problems efficiently while building compact and stable list of rules. SIRUS is built on random forests [13], and its general principle is as follows. Each node of each tree of a random forest can be transformed into an elementary rule. Hence, the core idea of SIRUS is to extract rules from a random forest based on their occurrence frequency. The most frequent rules, which stand for robust and strong data patterns, are eventually linearly combined to create the final predictions. The main competitors of SIRUS are RuleFit [42] and Node harvest [67]. Both methods work by extracting large collection of rules from tree aggregation: RuleFit is based on a boosted tree ensemble [41], whereas Node harvest is built on random forests. The rule selection procedure is implemented using a sparse linear aggregation, respectively the Lasso [93] for RuleFit and a constrained quadratic program for Node harvest. Yet, despite their excellent predictive performance, these two methods have a tendency to produce large, complex, and unstable lists of rules (usually of the order of 30–50), which questions their interpretability. Since tree ensemble methods are intrinsically random procedures, running the above algorithms several times on the same dataset outputs different lists of rules. SIRUS considerably improves stability and simplicity over its competitors, while preserving a comparable predictive accuracy and computational complexity. We first present SIRUS algorithm, and then analyze the associated theoretical properties, as well as the empirical performance on real data.

3.2.2.1   SIRUS Algorithm

SIRUS proceeds in the four following steps: the rule generation, the rule selection, the post-treatment of the rule set, and the rule aggregation.

Rule Generation SIRUS starts by training a random forest (with a large number M of trees) on the training set $\mathscr{D}_n$. The node splits are restricted to the q-empirical quantiles of the marginals $X^{(1)}, \ldots, X^{(p)}$ (with, by default, $q = 10$) in order to stabilize the forest structure. This minor modification of Breiman's original algorithm has a negligible impact on predictive accuracy, but is critical for stability (see Section 3 of the Supplementary Material in [5]). The resulting forest is broken down in a large collection of rules in the following way. Note that each (inner or terminal) node of each tree of the trained forest corresponds to a hyperrectangle in the input space $\mathbb{R}^p$. Consequently, each node can be transformed into an elementary regression rule, which is nothing but a piecewise constant estimate that takes two different values depending on whether the query point falls in the corresponding hyperrectangle or not. More precisely, each node is represented by a path, say $\mathscr{P}$, which enumerates the sequence of splits to reach the node from the root of the tree. In the following, we denote by Π the finite list of all possible paths, and stress out that each path $\mathscr{P} \in \Pi$ defines a regression rule. In the first step of SIRUS, both internal and external nodes are extracted from all trees of the random forest to generate a large collection of rules, typically of order 10^4.

Rule Selection Relevant rules are then selected among the large collection of rules created in the previous step. Indeed, despite the intrinsic tree randomization, there is some redundancy in the extracted rules, notably due to the quantile discretization implemented in the first step. Rules with high frequencies of appearance represent strong and robust data patterns, and are thus good candidates to compose a compact, stable, and predictive rule ensemble. For each possible path $\mathscr{P} \in \Pi$, we let $\hat{p}_{M,n}(\mathscr{P})$ be its appearance frequency. Then relevant rules are simply selected using a threshold $p_0 \in (0, 1)$, that is,

$$\hat{\mathscr{P}}_{M,n,p_0} = \{\mathscr{P} \in \Pi : \hat{p}_{M,n}(\mathscr{P}) > p_0\}.$$

The threshold p_0 is a tuning parameter. Its impact and optimal tuning will be discussed in the experimental section (Figs. 3.7 and 3.8). Optimal p_0 values essentially select rules made of one or two splits. Indeed, rules with a higher number of splits are more sensitive to data perturbation, hence associated to smaller frequencies $\hat{p}_{M,n}(\mathscr{P})$. Consequently, SIRUS builds shallow trees to decrease the computational cost while leaving the rule selection unaltered. In a nutshell, SIRUS uses randomized bagging, but aggregates the forest structure itself instead of its predictions, with the final aim of stabilizing the rule selection.

Rule Set Post-treatment Dependency between selected rules may occur due to the path extraction mechanism. For instance, consider the 6 rules extracted from a tree of depth 2: 2 rules are made of one split and 4 rules of two splits. Those 6 rules are linearly dependent since their associated hyperrectangles overlap. To properly define a linear aggregation of the selected rules, SIRUS filters, in a third step, $\hat{\mathscr{P}}_{M,n,p_0}$ with the following post-treatment mechanism: if the rule induced by the path $\mathscr{P} \in \hat{\mathscr{P}}_{M,n,p_0}$ is a linear combination of rules associated with paths with a higher appearance frequency, then $\mathscr{P}$ is removed from $\hat{\mathscr{P}}_{M,n,p_0}$.

Rule Aggregation So far, we have obtained a small set of regression rules. Recall that a rule $\hat{g}_{n,\mathscr{P}}$ corresponding to a path $\mathscr{P}$ is a piecewise constant estimate: if a query point $\mathbf{x}$ falls into the given hyperrectangle $H_{\mathscr{P}} \subset \mathbb{R}^p$, the rule returns the average of the Y_i's for the training points $\mathbf{X}_i$'s falling into $H_{\mathscr{P}}$; symmetrically, if $\mathbf{x}$ is located outside $H_{\mathscr{P}}$, the average of the Y_i's for training points that do not belong to $H_{\mathscr{P}}$ is output. Next, non-negative weights are assigned to the selected rules, in order to create one single estimate of the regression function $m(\mathbf{x})$. These weights are the solution of a ridge regression problem, in which each predictor is a rule $\hat{g}_{n,\mathscr{P}}$ for $\mathscr{P} \in \hat{\mathscr{P}}_{M,n,p_0}$ and weights are forced to be non-negative. The final aggregated estimate $\hat{m}_{M,n,p_0}(\mathbf{x})$ of $m(\mathbf{x})$ computed in the last step of SIRUS can be written as

$$\hat{m}_{M,n,p_0}(\mathbf{x}) = \hat{\beta}_0 + \sum_{\mathscr{P} \in \hat{\mathscr{P}}_{M,n,p_0}} \hat{\beta}_{n,\mathscr{P}} \hat{g}_{n,\mathscr{P}}(\mathbf{x}), \tag{3.1}$$

where $\hat{\beta}_0$ and $\hat{\beta}_{n,\mathscr{P}}$ are the ridge regression solutions. More precisely, denoting by $\hat{\beta}_{n,p_0}$ the column vector whose components are the coefficients $\hat{\beta}_{n,\mathscr{P}}$ for $\mathscr{P} \in \hat{\mathscr{P}}_{M,n,p_0}$, and letting $\mathbf{Y} = (Y_1, \ldots, Y_n)^T$ and $\boldsymbol{\Gamma}_{n,p_0}$ the matrix whose rows are the rule values $\hat{g}_{n,\mathscr{P}}(\mathbf{X}_i)$ for $i \in \{1, \ldots, n\}$, we have

$$(\hat{\boldsymbol{\beta}}_{n,p_0}, \hat{\beta}_0) = \underset{\boldsymbol{\beta} \geq 0, \beta_0}{\operatorname{argmin}} \frac{1}{n} \|\mathbf{Y} - \beta_0 \mathbf{1}_n - \boldsymbol{\Gamma}_{n,p_0} \boldsymbol{\beta}\|_2^2 + \lambda \|\boldsymbol{\beta}\|_2^2,$$

where $\mathbf{1}_n = (1, \ldots, 1)^T$ is the n-vector with all components equal to 1, and λ is a positive parameter tuned by cross-validation that controls the penalization strength. The infimum is taken over $\beta_0 \in \mathbb{R}$ and all the vectors $\boldsymbol{\beta} = \{\beta_1, \ldots, \beta_{c_n}\} \in \mathbb{R}_+^{c_n}$ with $c_n = |\hat{\mathscr{P}}_{M,n,p_0}|$ being the number of selected rules ($|\mathcal{A}|$ denotes the number of elements in any set $\mathcal{A}$). Besides, notice that the rule shape (with an else clause) differs from the classical format in the rule learning literature. This modification provides good properties of stability and modularity (investigation of the rules one by one [71]) to SIRUS—see Sect. 4 of the Supplementary Material in [5].

This aggregation procedure is a major step of SIRUS and deserves some additional comments. Indeed, in RuleFit, rules are also extracted from a tree ensemble, but aggregation is performed using Lasso procedure. By construction, the extracted rules are strongly correlated, and, unfortunately, the Lasso selection is known to be highly unstable in such correlated settings. This fact explains the instability of RuleFit, as demonstrated by the experiments in Sect. 3.2.2.3. On the other hand, the parameter p_0 controls the sparsity of SIRUS, and the ridge regression enables a stable rule aggregation. Furthermore, the constraint $\boldsymbol{\beta} \geq 0$ is added to force all coefficients to be non-negative, similarly to the Node harvest procedure [67]. Additionally, an unconstrained regression, in presence of correlated rules, would lead to negative values for some of the coefficients $\hat{\beta}_{n,\mathscr{P}}$. Such behavior would drastically undermines the algorithm interpretability.

Interpretability The notions of simplicity, stability, and predictivity, at the core of our definition of interpretability, need to be formally defined and quantified to enable comparison between algorithms. **Simplicity** refers to the model complexity, and particularly, to the number of operations involved in the prediction mechanism. In the specific case of rule algorithms, a natural measure of simplicity is provided by the number of rules. Furthermore, intuitively, a rule algorithm is **stable** when two independent estimations based on two independent samples return similar lists of rules. Formally, let $\hat{\mathscr{P}}'_{M,n,p_0}$ be the list of rules output by SIRUS fit on an independent sample $\mathscr{D}'_n$. Then the proportion of rules shared by $\hat{\mathscr{P}}_{M,n,p_0}$ and $\hat{\mathscr{P}}'_{M,n,p_0}$ gives a stability measure. Such a metric is known as the Dice-Sorensen index, and is often used to assess variable selection procedures [2, 9, 19, 50, 103]. In our case, the Dice-Sorensen index is then defined as

$$\hat{S}_{M,n,p_0} = \frac{2|\hat{\mathscr{P}}_{M,n,p_0} \cap \hat{\mathscr{P}}'_{M,n,p_0}|}{|\hat{\mathscr{P}}_{M,n,p_0}| + |\hat{\mathscr{P}}'_{M,n,p_0}|}.$$

Unfortunately, in practice one rarely has access to an additional sample $\mathscr{D}_n'$. Therefore, to tackle this issue, we use a 10-fold cross-validation to simulate data perturbation. The stability metric is thus empirically defined as the average proportion of rules shared by two models of two distinct folds of the cross-validation. A stability of 1 means that the exact same list of rules is selected over the 10 folds, whereas a stability of 0 means that all rules are distinct between any 2 folds. Regarding **predictivity**, a natural measure of prediction error in regression problems is given by the proportion of unexplained variance. We use a 10-fold cross-validation to estimate this proportion.

3.2.2.2 Theoretical Analysis

Stability is the most critical of the three minimum requirements for obtaining interpretable models. In SIRUS, simplicity is controlled by the hyperparameter p_0. Existing works on rule learning provide many experiments highlighting that rule algorithms and tree ensemble share similar accuracy performances. On the other hand, creating a stable rule procedure is more challenging [60, 71]. Consequently, we focus our theoretical analysis on the asymptotic stability of SIRUS.

First, we need a rigorous definition of the rule extraction procedure. We introduce a symbolic representation of a path in a tree, which describes the sequence of splits to reach a given (inner or terminal) node from the root. Note that such path encoding can be used in both the empirical and theoretical algorithms to define rules. We define a path $\mathscr{P}$ as

$$\mathscr{P} = \{(j_k, r_k, s_k), \ k = 1, \ldots, d\},$$

where d is the tree depth, and for $k \in \{1, \ldots, d\}$, the triplet (j_k, r_k, s_k) describes how to move from level $(k-1)$ to level k, with a split using the coordinate $j_k \in \{1, \ldots, p\}$, the index $r_k \in \{1, \ldots, q-1\}$ of the corresponding quantile, and a side $s_k = L$ if we go to the left and $s_k = R$ if we go to the right (see Fig. 3.5). We denote by $\varPi$ the set of all possible such paths.

Each tree of the forest is randomized in two ways: (i) a bootstrap sample based on the data set $\mathscr{D}_n$ is created prior to the tree construction, and (ii) the best split at each node is selected along a randomly selected subset of variables. These two randomization mechanisms are governed by a random variable called Θ. We let $T(\Theta, \mathscr{D}_n)$ be a random subset of $\varPi$, consisting of the collection of the extracted paths from the random tree built with the randomness Θ and the dataset $\mathscr{D}_n$. Now, let $\Theta_1, \ldots, \Theta_M$ be M independent random variables that are used to build the M different trees in the forest. The empirical frequency of occurrence of a path $\mathscr{P} \in \varPi$ in the forest can be written as

$$\hat{p}_{M,n}(\mathscr{P}) = \frac{1}{M} \sum_{\ell=1}^{M} \mathbb{1}_{\mathscr{P} \in T(\Theta_\ell, \mathscr{D}_n)},$$

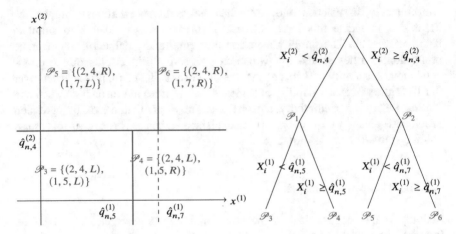

Fig. 3.5 Example of a partition of a tree of depth 2: the tree on the right, the associated paths and hyperrectangles of length $d = 2$ on the left

which is nothing but the proportion of trees that contain $\mathscr{P}$. By definition, $\hat{p}_{M,n}(\mathscr{P})$ is the Monte Carlo estimate of the probability $p_n(\mathscr{P})$ that a Θ-random tree contains a particular path $\mathscr{P} \in \Pi$, that is,

$$p_n(\mathscr{P}) = \mathbb{P}(\mathscr{P} \in T(\Theta, \mathscr{D}_n)|\mathscr{D}_n).$$

Now, we introduce all theoretical counterparts of the previous empirical quantities, which, by definition, do not depend on the sample $\mathscr{D}_n$ but on the unknown distribution of $(\mathbf{X}, Y)$. We denote by $T^\star(\Theta)$ the list of all paths contained in the theoretical tree built with the extra randomness Θ, in which splits are selected in order to maximize the population CART-splitting criterion instead of the empirical one. The probability $p^\star(\mathscr{P})$ that a given path $\mathscr{P}$ belongs to a theoretical randomized tree (which is nothing but the theoretical counterpart of $p_n(\mathscr{P})$) is

$$p^\star(\mathscr{P}) = \mathbb{P}(\mathscr{P} \in T^\star(\Theta)).$$

Lastly, we let $\mathscr{P}^\star_{p_0} = \{\mathscr{P} \in \Pi : p^\star(\mathscr{P}) > p_0\}$ be the theoretical set of selected paths, with the same post-treatment as for the data-based procedure (removing linear dependencies and discarding paths with a null coefficient in the rule aggregation). Similarly to previous theoretical works on random forests [69, 82], we assume that the subsampling of a_n observations prior to each tree construction is done without replacement to ease the mathematical analysis. Our stability result holds under the following mild assumptions:

(A1) The subsampling rate a_n satisfies $\lim_{n \to \infty} a_n = \infty$ and $\lim_{n \to \infty} \frac{a_n}{n} = 0$, and the number of trees M_n satisfies $\lim_{n \to \infty} M_n = \infty$. $\square$

(A2) The random variable $\mathbf{X}$ has a strictly positive density f with respect to the Lebesgue measure on $\mathbb{R}^p$. Furthermore, for all $j \in \{1, \ldots, p\}$, the marginal density $f^{(j)}$ of $X^{(j)}$ is continuous, bounded, and strictly positive. Finally, the random variable Y is bounded. □

Theorem 3.1 *Grant Assumptions (A3.2.2.2) and (A3.2.2.2). Let $\mathcal{U}^\star = \{p^\star(\mathscr{P}) : \mathscr{P} \in \Pi\}$ be the set of all theoretical probabilities of appearance for each path $\mathscr{P}$. Then, provided $p_0 \in [0, 1] \setminus \mathcal{U}^\star$ and $\lambda > 0$, we have*

$$\lim_{n \to \infty} \hat{S}_{M_n, n, p_0} = 1 \quad \text{in probability.}$$

Theorem 3.1 establishes the stability of SIRUS: assuming that the sample size is large enough, the same list of rules is output across several fits on independent samples. An inspection of the proof of Theorem 3.1 reveals that the cut discretization (performed at quantile values only), as well as considering random forests (instead of boosted tree ensembles as in RuleFit) are the crucial features that allow us to stabilize rule models extracted from tree ensembles. The experiments in Sect. 3.2.2.3 also highlight the high empirical stability of SIRUS in finite-sample regimes.

3.2.2.3 Experiments

We describe here the experiments presented in [5] to demonstrate the improvement of SIRUS over state-of-the-art methods.[1]

SIRUS Rule Set SIRUS is illustrated on the "LA Ozone" dataset from Friedman et al. [40], which records the level of atmospheric ozone concentration from eight daily meteorological measurements made in Los Angeles in 1976: wind speed ("wind"), humidity ("humidity"), temperature ("temp"), inversion base height ("ibh"), daggot pressure gradient ("dpg"), inversion base temperature ("ibt"), visibility ("vis"), and day of the year ("doy"). The response "Ozone" is the log of the daily maximum of ozone concentration. The list of rules output for this dataset is presented in Fig. 3.6. The column "Frequency" refers to $\hat{p}_{M,n}(\mathscr{P})$, the occurrence frequency of each rule in the forest, used for rule selection. Figure 3.6 enables us to understand how weather conditions influence the ozone concentration. More specifically, a temperature larger than 65°F or a high inversion base temperature results in high ozone concentrations. The third rule in Fig. 3.6 shows that the interaction of a high temperature with a visibility lower than 150 miles induces even higher levels of ozone concentration. Strikingly, the ninth rule reveals that especially low ozone concentrations are reached when a low temperature and a low inversion base temperature are combined. Recall that to generate a prediction for a

---

[1] See Table 1 in Sect. 5 of the Supplementary Material in [5] for dataset details.

Average Ozone = 12			Intercept = −7.8				
Frequency	**Rule**						**Weight**
0.29	if	temp < 65	then	Ozone = 7	else	Ozone = 19	0.12
0.17	if	ibt < 189	then	Ozone = 7	else	Ozone = 18	0.07
0.063	if	temp ≥ 65 and vis < 150	then	Ozone = 20	else	Ozone = 7	0.31
0.061	if	vh < 5840	then	Ozone = 10	else	Ozone = 20	0.072
0.060	if	ibh < 2110	then	Ozone = 16	else	Ozone = 7	0.14
0.058	if	ibh < 2960	then	Ozone = 15	else	Ozone = 6	0.10
0.051	if	temp ≥ 65 and ibh < 2110	then	Ozone = 21	else	Ozone = 8	0.16
0.048	if	vis < 150	then	Ozone = 14	else	Ozone = 7	0.18
0.043	if	temp < 65 and ibt < 120	then	Ozone = 5	else	Ozone = 15	0.15
0.040	if	temp < 70	then	Ozone = 8	else	Ozone = 20	0.14
0.039	if	ibt < 227	then	Ozone = 9	else	Ozone = 22	0.21

Fig. 3.6 SIRUS rule list for the "LA Ozone" dataset

given query point **x**, for each rule the corresponding ozone concentration is retrieved depending on whether **x** satisfies the rule conditions. Then all rule outputs for **x** are multiplied by their associated weight and added together. One can notice that rule importances and weights are not related. For instance, the third rule has a larger weight than the most two important ones. It is obvious that rule 3 has multiple constraints and is consequently more sensitive to data perturbation—hence a smaller frequency of appearance in the forest. On the other hand, its associated variance decrease in CART is more important than for the first two rules, leading to a larger weight in the linear combination. Since rules 5 and 6 are strongly correlated, their weights are diluted.

Tuning SIRUS has only one hyperparameter which requires fine tuning: the threshold p_0 to control the model size by selecting the most frequent rules in the forest. First, the range of possible values of p_0 is defined so that the model size varies between 1 and 25 rules. This arbitrary upper bound is a safeguard to avoid complex and long list of rules that are difficult to interpret properly. In practice, this limit of 25 rules is rarely reached, since the following tuning of p_0 often leads to compact rule lists. Next, p_0 is tuned within that range by cross-validation to maximize both stability and predictivity. To find a tradeoff between these two properties, we follow a standard bi-objective optimization procedure:[2] p_0 is chosen to be as close as possible to the ideal case of 0 unexplained variance and 90% stability. This tuning procedure is computationally fast: the cost of about 10 fits of SIRUS. Besides, the

---

[2] See Sect. 2 of the Supplementary Material in [5] for details on the bi-objective procedure.

optimal number of trees M is set automatically by SIRUS: as stability, predictivity, and computation time increase with the number of trees, no fine tuning is required for M. Thus, a stopping criterion is designed to grow the minimum number of trees which enforces that stability and predictivity are greater than 95% of their maximum values (reached when M tends to infinity).[3] Finally, we use the other standard settings of random forests (well-known for their excellent performance), set $q = 10$ quantiles, and transform categorical variables into multiple binary variables.

Performance We compare SIRUS performance with that of its two main competitors RuleFit (with rule predictors only) and Node harvest. For predictive accuracy, we provide baselines by running random forests and (pruned) CART. In order to compute stability metrics only, data is binned using 10 quantiles to fit Rulefit and Node harvest. Our R/C++ package `sirus` is adapted from `ranger`, a fast random forests implementation [99]. We also use available R implementations `pre` [36, RuleFit] and `nodeharvest` [68]. While SIRUS predictive performance is comparable to that of Node harvest and slightly below RuleFit, its stability is considerably improved with much smaller rule lists. Experimental results are gathered in Table 3.1a for model sizes, Table 3.1b for stability, and Table 3.2 for predictive accuracy. All results are averaged over 10 repetitions of the cross-validation procedure. To increase readability, and because they are negligible, standard deviations are not displayed. In the last column of Table 3.2, we choose p_0 so that the number of rules in SIRUS is equal to that of RuleFit and Node harvest (about 50 rules): in this setting, SIRUS predictivity is then as good as that of RuleFit.

To comment the behavior of our method, we illustrate our results with two specific datasets: "Diabetes" [33] and "Machine" [32]. The "Diabetes" data contains $n = 442$ diabetic patients and the response of interest Y is a measure of disease progression over one year. A total of 10 variables are collected for each patient: age, sex, body mass index, average blood pressure, and six blood serum measurements $s1, s2, \ldots, s6$. For this dataset, SIRUS is as predictive as a random forest, with only 12 rules when the forest performs about 10^4 operations: the unexplained variance is 0.56 for SIRUS and 0.55 for random forest. Notice that CART performs considerably worse with 0.67 unexplained variance. For the "Machine" dataset, the output Y of interest is the CPU performance of computer hardware. For $n = 209$ machines, 7 variables are collected about the machine characteristics. For this dataset, SIRUS, RuleFit, and Node harvest have a similar predictivity, in-between CART and random forests. Our algorithm achieves such performance with a readable list of only 9 rules stable at 88%, while RuleFit and Node harvest incorporate respectively 44 and 42 rules with stability levels of 23% and 29%. Stability and predictivity are represented as p_0 varies for "Diabetes" and "Machine" datasets in Figs. 3.7 and 3.8, respectively.

---

[3] See Sect. 6 of the Supplementary Material in [5] for a detailed definition of this criterion.

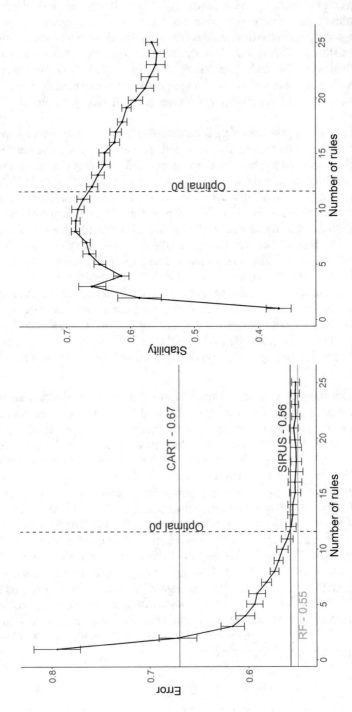

Fig. 3.7 For the dataset "Diabetes", unexplained variance (left panel) and stability (right panel) versus the number of rules when p_0 varies, estimated via 10-fold cross-validation (results are averaged over 10 repetitions)

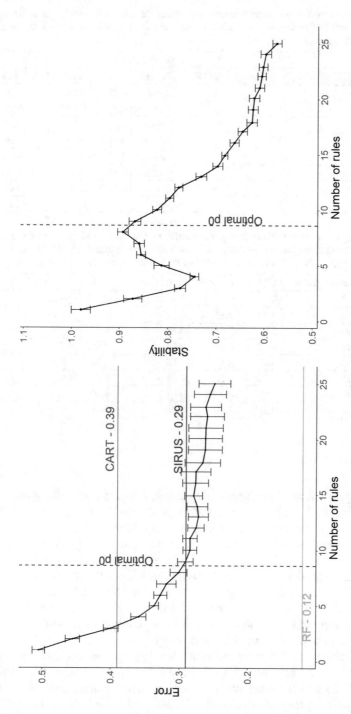

Fig. 3.8 For the dataset "Machine", unexplained variance (left panel) and stability (right panel) versus the number of rules when p_0 varies, estimated via 10-fold cross-validation (results are averaged over 10 repetitions)

Table 3.1 Mean model size and stability over a 10-fold cross-validation for several public datasets. In Table a, the smallest model size for each dataset is displayed in bold. In Table b, the highest stability for each dataset is in bold

(a) Model size

Dataset	CART	RuleFit	Node harvest	SIRUS
Ozone	15	21	46	**11**
Mpg	15	40	43	**9**
Prostate	**11**	14	41	23
Housing	15	54	40	**6**
Diabetes	**12**	25	42	**12**
Machine	**8**	44	42	9
Abalone	20	58	35	**6**
Bones	17	5	13	**1**

(b) Stability

Dataset	RuleFit	Node harvest	SIRUS
Ozone	0.22	0.30	**0.62**
Mpg	0.25	0.43	**0.83**
Prostate	0.32	0.23	**0.48**
Housing	0.19	0.40	**0.80**
Diabetes	0.18	0.39	**0.66**
Machine	0.23	0.29	**0.88**
Abalone	0.31	0.38	**0.82**
Bones	0.59	0.52	**0.89**

Table 3.2 Proportion of unexplained variance estimated over a 10-fold cross-validation for several public datasets. For rule algorithms only, i.e., RuleFit, Node harvest, and SIRUS, maximum values are displayed in bold, as well as values within 10% of the maximum for each dataset.

Dataset	Random forest	CART	RuleFit	Node harvest	SIRUS	SIRUS 50 Rules
Ozone	0.25	0.36	**0.27**	0.31	0.32	**0.26**
Mpg	0.13	0.20	**0.15**	0.20	0.21	**0.15**
Prostate	0.48	0.60	**0.53**	**0.52**	**0.48**	0.55
Housing	0.13	0.28	**0.16**	0.24	0.31	0.21
Diabetes	0.55	0.67	**0.55**	**0.58**	**0.56**	**0.54**
Machine	0.13	0.39	**0.26**	**0.29**	**0.29**	**0.27**
Abalone	0.44	0.56	**0.46**	0.61	0.66	0.64
Bones	0.67	0.67	**0.70**	**0.70**	**0.74**	**0.72**

3.2.3 Discussion

Rule learning models are very attractive approaches when interpretability is required because of their high simplicity and excellent predictive skills when data exhibits nonlinear patterns. However, rule models are also known to be often strongly unstable, as most interpretable models such as decision trees or GAM. Although a large number of rule algorithms have been developed since the 1980s, very few works have focused on this stability issue. To address this problem, we have designed SIRUS, a stable rule model for regression. Note that SIRUS is also available for classification [7]. We proved the theoretical stability of SIRUS and extensive experiments showed the high stability improvement over state-of-the-art algorithms, while preserving simplicity and accuracy.

Clearly, simplicity and accuracy are contradictory properties as a higher number of rules improves the accuracy of a rule model. On the other hand, the relation between stability and the other two properties is more complex and interesting. Indeed, compared to the main competitors RuleFit and Node harvest, the stability mechanism developed in SIRUS naturally leads to shorter rule lists for a comparable accuracy. Additionally, when the number of rules varies, a stability peak of SIRUS often coincides with an area where the accuracy does not vary much, and is quite close to the asymptotic maximum. These two remarkable behaviors have motivated the tuning procedures of SIRUS, and it could be interesting to deepen their analysis through further experiments.

It is also worth noticing that SIRUS stability is a result of the simple trees used in the forest: the splits occurred at the quantile levels along each marginal and the tree depth is limited to two. These choices have been made to restrict the number of eligible splits, with the final aim to detect second-order interactions. If the underlying phenomenon generating the data involves higher-order interactions, SIRUS accuracy could be drastically lower than that of random forests. In our experiments, the decrease of accuracy of SIRUS is surprisingly small with respect to random forests, especially regarding that SIRUS usually outputs about only ten rules. The following fact may explain this good accuracy: a rule model is a piecewise constant estimate, and since an input observation can satisfy any rule subset, ten rules generates an estimate made of $2^{10} = 1024$ pieces. Such sharp decomposition of the input space may explain the high accuracy of SIRUS despite the small number of rules. Obviously, if the data is complex enough and the sample is large enough, we may observe a significant gap of accuracy between SIRUS and random forests. It would be interesting to deepen experiments to better understand in which cases the gap of accuracy between SIRUS and random forests becomes large. Such analyses may provide insights to improve the efficiency of rule extraction from tree ensembles.

3.3   Post-Processing of Black-Box Algorithms via Variable Importance

We have seen with SIRUS that it is possible to design interpretable models that are simple, accurate and stable. Another approach toward interpretability uses black-box algorithms (as random forests or neural networks) instead of interpretable models, and tries to understand *a posteriori* their behavior. Variable importance analysis is the most widely used post-hoc approach to interpret statistical learning algorithms. The goal is to rank all input variables by decreasing order of influence in the prediction process of the black-box model. As highlighted by Genuer et al. [45], there are essentially two final objectives of a variable importance analysis. A first goal is to reduce the problem dimension by the selection of a small number of input variables with a maximized predictive accuracy. A second objective is to detect and rank all influential variables to focus on for further exploration with domain experts. We illustrate the difference between these two goals by considering the case of two highly correlated and influential variables. Since these two variables contain the same information, one should be removed for the first objective, whereas both should be kept for the second objective as they may represent distinct quantities associated to different interpretations for domain experts.

Many supervised learning methods have specific global variable importance measures, typically linear models, tree ensembles, or neural networks. Some other importance measures are model agnostic since they can be applied to any black-box predictor—permutation measures for example. Also notice that variable importance measures often have an empirical definition, as opposed to most other post-hoc methods, which first define the theoretical targeted quantities and build estimates in a second step. We review the existing variable importance measures in Sect. 3.3.1 and their connections with sensitivity analysis, which will be helpful to understand the asymptotic behavior of these indicators. In Sect. 3.3.2, we study the asymptotic behavior of the MDA (Mean Decrease Accuracy), one of the two variable importance measures used in random forests. This analysis reveals that this indicator is asymptotically biased. We therefore design a new indicator, called Sobol-MDA, and proves its accuracy from a theoretical and empirical point of view. Finally, Sect. 3.3.3 proposes an efficient way to compute Shapley effects, which is an importance measure useful in presence of dependence and interactions.

3.3.1 Literature Review

3.3.1.1 Model-Specific Variable Importance

Tree Ensembles There are essentially two measures of variable importance for tree ensembles. The first one is the Mean Decrease Accuracy (MDA), defined by Breiman [13] for random forests: the values of a given variable are permuted, then predictions are computed for these perturbed data points with the corresponding accuracy. The difference between this degraded accuracy and the original one gives the importance of the variable. A second approach is the Mean Decrease Impurity (MDI), based on the total decrease in node impurity from splitting on a given variable in a single tree. The MDI is defined for all kind of tree ensembles: the tree impurity decrease is averaged over all trees for random forests [15], whereas it is summed across all boosting iterations for boosted ensembles [21, 39]. These two measures behave poorly when the correlation within input variables is high [3, 90]. Gregorutti et al. [47] alleviate this issue by combining random forests and the MDA with the Recursive Feature Elimination (RFE) algorithm to perform backward variable selection.

Secondly, several algorithms tackle the problem of detecting high-order interactions in tree ensembles, initially Random Intersection Trees RIT [83], and more recently iterative random forests iRF [4] that combines ideas from RIT and RFE. Signed iterative Random Forests [58] enriches high-order feature interactions with a thresholding behavior for each variable, to indicate if rather low or high values are of interest.

Finally, since variable importance measures are quite strongly biased when input variables are correlated, several approaches were recently developed to improve importance measures in such a setting [18, 69]. The main principle is to retrain the learning algorithm by removing variables from the training data, and compare if predictions differ from the original model with all input variables. Mentch and Hooker [69] tackle the specific case of random forests. Proving the asymptotic normality of forest predictions, they design statistical tests to detect if predictions are significantly modified by the removal of given input variables. On the other hand, the approach of Candes et al. [18] is model agnostic. The idea is to add noisy variables that are independent of the output conditional on the other inputs to detect if the importance measures of the original input variables are significant.

Neural Networks Global variable importance has received less attention for neural networks than for tree ensembles, as opposed to local importance measures as we will see in the following subsection. This fact has a straightforward explanation. Indeed, neural networks are mainly applied to data with spatial structures, typically images, where the global importance of a fixed pixel over the full training data is not really meaningful. For example, to predict the presence of a given object in an image, the importance of a fixed pixel completely relies on the object position. On the other hand, a local importance measure identifies the image areas responsible for the prediction, and is clearly a more relevant approach than global measures. However, we can mention a few global importance measures for neural networks, with the approach from Erhan et al. [34], which identifies inputs maximizing the activation of each layer of the neural network. More recently, Ish-Horowicz et al. [53] extend RATE [27] to the Bayesian deep learning setting. The main idea is to compute the projection of the function learned by the network onto the input observation matrix, where such a projection is called the effect size analog $\tilde{\beta}$. If $\tilde{\beta}^{(-j)}$ is the vector $\tilde{\beta}$ without the j-th component, RATE is defined as the Kullback-Leibler divergence of the distribution of $\tilde{\beta}^{(-j)}$ and the distribution of $\tilde{\beta}^{(-j)}$ conditional on $\tilde{\beta}^{(j)}$, which provides a global importance measure for $X^{(j)}$. Finally, Kim et al. [56] design an interesting approach for global interpretations of neural networks with Concept Activation Vectors (CAV). Instead of estimating an importance measure for each input, the goal is to identify the images associated to a given concept. Figure 2 in Kim et al. [56] provides an example of CAV, where the left panel displays images of stripes which are the most and the least related to the "CEO" concept. The most similar images are pinstripes, typically related to the suit or shirt of a CEO.

3.3.1.2 Global Sensitivity Analysis

Global sensitivity analysis (GSA) is the study of uncertainties in a system. In particular, the main goal of GSA is to determine how the uncertainty in the model output is apportioned to the uncertainty of the different inputs. Such analyses enable us to identify variables that strongly influence the output, and those with

no influence. For detailed reviews of GSA, we refer to Iooss and Lemaître [51] and
Ghanem et al. [46]. Thus, sensitivity analysis is close to variable importance for
learning algorithms. However, this last type of methods usually have an algorithmic
definition, as opposed to sensitivity analysis, where importance measures are
first formally defined based on the data distribution. Then, in a second step,
the theoretical quantities are estimated, usually using models and Monte-Carlo
methods. One of the main importance measures is Sobol indices based on variance
decomposition [81, 87], and variances of the output expectation conditional on
subsets of input variables. It enables us to quantify the importance of each input
variable as well as their interactions for any black-box model when inputs are
independent. However, in the dependent case, the interpretation of Sobol indices
becomes difficult. Instead, we rather use Shapley effects in such settings, which
equitably allocate the output variance due to dependence and interactions across all
input variables [52, 73, 88].

Sobol Indices Multiple Sobol indices exist to measure the main effect or the total
effect of a given variable, as well as the interaction between two variables [81, 87].
We first consider the case where $X^{(1)}, \ldots, X^{(p)}$ are independent, which enables
a clear interpretation of Sobol indices. For $j \in \{1, \ldots, p\}$, the first order Sobol
index $S^{(j)}$ measures the impact on the output of a given variable $X^{(j)}$ alone, and is
formally defined as

$$S^{(j)} = \frac{\mathbb{V}[\mathbb{E}[Y|X^{(j)}]]}{\mathbb{V}(Y)}.$$

Next, the Sobol index measures the total contribution of $X^{(j)}$ to the output variance,
including the interactions of $X^{(j)}$ with all other variables, that is

$$ST^{(j)} = \frac{\mathbb{E}[\mathbb{V}[m(\mathbf{X})|\mathbf{X}^{(-j)}]]}{\mathbb{V}(Y)},$$

where $\mathbf{X}^{(-j)}$ is the random vector $\mathbf{X}$ without the j-th component, and $m(\mathbf{x}) = \mathbb{E}[Y|\mathbf{X} = \mathbf{x}]$ is the regression function. We also define second order Sobol indices
$S^{(j,k)}$ for $k \in \{1, \ldots, p\}$ as

$$S^{(j,k)} = \frac{\mathbb{V}[\mathbb{E}[Y|X^{(j)}, X^{(k)}]]}{\mathbb{V}(Y)} - S^{(j)} - S^{(k)},$$

which measure the contribution of the interaction of $X^{(j)}$ and $X^{(k)}$ to the output
variance. It is obviously possible to extend such definition to higher-order Sobol
indices. From the ANOVA decomposition [20, 87], we have

$$\sum_{j=1}^{p} S^{(j)} + \sum_{j,k} S^{(j,k)} + \sum_{j,k,\ell} S^{(j,k,\ell)} + \ldots = \sum_{U \subset \{1,\ldots,p\}} S^{(U)} = 1, \qquad (3.2)$$

and

$$ST^{(j)} = S^{(j)} + \sum_{k=1}^{p} S^{(j,k)} + \sum_{k,\ell} S^{(j,k,\ell)} + \ldots = \sum_{U \subset \{1,\ldots,p\}\setminus\{j\}} S^{(U \cup \{j\})}.$$

When input variables are dependent, the ANOVA decomposition does not hold anymore in general, and we therefore loose the properties of Equation (3.2) which states that the sum of Sobol indices of all orders is 1. It is also not possible to separate contributions due to interactions from dependence, and higher-order Sobol indices become meaningless. However, total Sobol indices preserve a useful interpretation in the dependent setting. Indeed, the total Sobol index of variable $X^{(j)}$ gives the proportion of output variance lost when $X^{(j)}$ is removed from the model.

Shapley Effects In the case where inputs are dependent, Shapley effects are rather used instead of Sobol indices, as they equitably allocate the mutual contributions due to dependence and interactions to each individual input [52, 73, 88]. Shapley values were initially defined in economics and game theory [84] to solve the problem of attributing the value produced by a joint team to its individual members. The main idea behind Shapley values is to measure the difference of produced value between a subset of the team and the same subteam with an additional member. For a specific member, this difference is averaged over all possible subteams and gives his Shapley value. Recently, Owen [73] adapted Shapley values to the problem of variable importance in statistical learning, where an input variable plays the role of a member of the team, and the produced value is the explained output variance. In this statistical learning context, Shapley values are now called Shapley effects, and are extensively used to interpret both tree ensembles and neural networks. To formalize Shapley effects, we denote by $\mathbf{X}^{(U)}$ the subvector with only the components in $U \subset \{1, \ldots, p\}$. Then, the Shapley effect of the j-th variable is defined by

$$Sh^{(j)} = \sum_{U \subset \{1,\ldots,p\}\setminus\{j\}} \frac{1}{p} \binom{p-1}{|U|}^{-1} \frac{\mathbb{V}[\mathbb{E}[Y|\mathbf{X}^{(U \cup \{j\})}]] - \mathbb{V}[\mathbb{E}[Y|\mathbf{X}^{(U)}]]}{\mathbb{V}[Y]}.$$

To put if differently, the Shapley effect of $X^{(j)}$ is the additional output explained variance when j is added to a subset $U \subset \{1, \ldots, p\}$, averaged over all possible subsets. The variance difference is averaged for a given size of U through the combinatorial weight, and then the average is taken over all U sizes through the term $1/p$. Note that the sum has 2^{p-1} terms, and each of them requires to estimate $\mathbb{V}[\mathbb{E}[Y|\mathbf{X}^{(U)}]]$, which is computationally costly and difficult to estimate accurately.

In the literature, efficient strategies have been designed to tackle these two issues. They all have drawbacks: they are either fast but with limited predictive performances, or accurate but computationally costly. In general, the computational issue of Shapley algorithm is addressed using Monte-Carlo methods [26, 88, 98]. Regarding the second issue of estimating conditional expectations, two main approaches exist: training one model for each selected subset of variables (accurate but not computationally efficient) [98], or training a single model once with all input variables and use greedy heuristics to derive the conditional expectations (computationally efficient but with limited accuracy). In the latter case, existing algorithms estimate the conditional expectations with a strong bias when input variables are dependent. More precisely, SAGE approach [26] simply replaces the conditional expectations by the marginal distributions, and Broto et al. [17] leverage k-nearest neighbors to approximate sampling from the conditional distributions. Besides, efficient algorithms exist in the specific setting where it is possible to draw samples from the conditional distributions of the inputs [1, 17, 88].

3.3.1.3 Local Interpretability

Local interpretations focus on the explanation of a single prediction. Indeed, variable importance measures can be adapted locally to give the contribution of each input variable to a given prediction.

SHAP Values Shapley effects are naturally adapted by Lundberg and Lee [64] to local importance measures, called SHAP values, by replacing the value function as follows:

$$\text{SHAP}^{(j)}(\mathbf{x}) = \sum_{U \subset \{1,\dots,p\}\backslash\{j\}} \frac{1}{p}\binom{p-1}{|U|}^{-1} \left(\mathbb{E}[Y|\mathbf{X}^{(U \cup \{j\})} = \mathbf{x}^{(U \cup \{j\})}] - \mathbb{E}[Y|\mathbf{X}^{(U)} = \mathbf{x}^{(U)}]\right),$$

which establishes how the prediction at the input point $\mathbf{x}$ is shifted by variable $X^{(j)}$ towards higher or lower values. This is illustrated in Fig. 3.9, where the considered input point is displayed on the vertical axis along variables, and blue contributions indicates that the variable reduces the prediction value, whereas red contributions indicate an increase of the prediction. Several algorithms were developed to estimate SHAP values. Initially, the KernelSHAP method [64] introduces an efficient trick to estimate SHAP values by solving a least-square regression problem. Indeed, if $I(U)$ is the binary vector of dimension p where the j-th component takes the value 1 if $j \in U$ and 0 otherwise, SHAP values are the minimum in β of the following cost function:

$$\ell(\beta, \mathbf{x}) = \sum_{U \subset \{1,\dots,p\}} w(U)(\mathbb{E}[Y|\mathbf{X}^{(U)} = \mathbf{x}^{(U)}] - \beta^T I(U))^2,$$

where the weights $w(U)$ are given by

$$w(U) = \frac{p-1}{\binom{p}{|U|}|U|(p-|U|)},$$

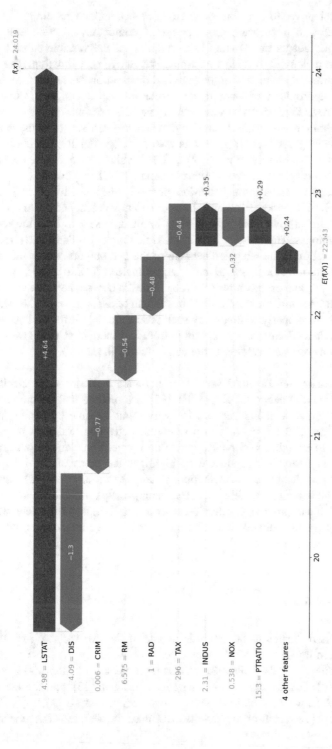

Fig. 3.9 SHAP values for the UCI dataset "Boston Housing", generated using the python package SHAP [64]

and the coefficient vector β is constrained to have its components sum to $\mathbb{E}[Y|\mathbf{X} = \mathbf{x}]$. Additionally, to circumvent the exponential computational complexity with p, KernelSHAP adapts the Monte-Carlo sampling of the variable subsets $U \subset \{1, \ldots, p\}$ from Song et al. [88] to estimate the above cost function. The value function is estimated simply using the marginal distribution of the inputs, which is a quite strong approximation when input variables are dependent. Next, Covert and Lee [25] improve KernelSHAP by mitigating the bias and introducing a variance reduction technique with paired sampling: when a given subset U is sampled, the complementary set $\{1, \ldots, p\} \setminus U$ is also selected. Finally, Lundberg et al. [65] introduce a fast algorithm to compute SHAP values for tree ensembles. The principle is to modify the tree predictions to estimate $\mathbb{E}[Y|\mathbf{X}^{(U)}]$ instead of $\mathbb{E}[Y|\mathbf{X} = \mathbf{x}]$, leaving the initial trees untouched. More precisely, the recursive algorithm from Lundberg et al. [65] works as follows: the query point $\mathbf{x}$ is dropped down each tree, but when a split on a variable outside of U is hit, $\mathbf{x}$ is sent to both the left and right children nodes. Therefore, $\mathbf{x}$ falls in multiple terminal cells of each tree. The final tree prediction is the weighted average of the cell outputs, where the weight associated to a terminal leaf A is given by an estimate of $\mathbb{P}(\mathbf{X} \in A|\mathbf{X}^{(U)} = \mathbf{x}^{(U)})$, defined as the product of the empirical probabilities to choose the side that leads to A at each split on a variable outside of U in the path of the original tree. Notice that these weights are properly estimated by such procedure only if the components of $\mathbf{X}$ are independent. Therefore, the algorithm from Lundberg et al. [65] gives biased predictions in a correlated setting, as noticed in Aas et al. [1].

Neural Networks Several local variable importance measures were specifically developed for neural networks. DeepLIFT [85] is a method that decomposes the output prediction of a neural network to every input variable by comparing for each neuron the actual activation to a reference activation. Saliency maps [86] are a method to explain the classification of an image. The class output gradient is computed at a given input image to highlight the areas of the image that are discriminative for the class prediction. Finally, Vaswani et al. [95] introduce attention methods for neural networks. The main principle is to learn an attention function, which can provide an attention score for each input of the network. For example in image recognition, the influence of each image area on the prediction can be computed.

3.3.2   Sobol-MDA

This section is dedicated to the analysis of one of the two variable importances used in random forests: the MDA (Mean Decrease Accuracy). A close inspection of the literature about the MDA shows that three versions of the MDA coexist: the Train-Test MDA, the Breiman-Cutler MDA Breiman [13], and the Ishwaran-Kogalur MDA Ishwaran et al. [55]. The last two versions improves over the Train-Test MDA by using the out-of-bag sample to estimate the decrease of accuracy when a given variable is noise up by permutation. The Breiman-Cutler MDA is a Monte-Carlo estimate of the tree decrease of accuracy, whereas the Ishwaran-Kogalur MDA measures the forest error decrease. Bénard et al. [8] conducted a theoretical analysis of the MDA, which shows that all MDA versions are asymptotically biased when variables are dependent, as formally stated in Proposition 3.1 below. Indeed, it is possible to break down the MDA limits using total Sobol indices, $ST^{(j)}$ defined above and $ST_{mg}^{(j)}$ (the total Sobol index computed for $\mathbf{X}_{\pi_j}$, the vector $\mathbf{X}$ where the j-th component is replaced by an independent copy of $X^{(j)}$), and another quantity named $\text{MDA}_3^{\star(j)}$, which is not an importance measure.[4]

Proposition 3.1 ([8]) *Under classical assumptions on the regression model and a good choice of the random forest parameters, assuming that theoretical trees are consistent, for all $M \in \mathbb{N}^\star$ and $j \in \{1, \ldots, p\}$ we have*

$$(i) \quad \widehat{\text{MDA}}_{M,n}^{(TT)}(X^{(j)}) \xrightarrow{\mathbb{L}^1} \mathbb{V}[Y] \times ST^{(j)} + \mathbb{V}[Y] \times ST_{mg}^{(j)} + \text{MDA}_3^{\star(j)}$$

$$(ii) \quad \widehat{\text{MDA}}_{M,n}^{(BC)}(X^{(j)}) \xrightarrow{\mathbb{L}^1} \mathbb{V}[Y] \times ST^{(j)} + \mathbb{V}[Y] \times ST_{mg}^{(j)} + \text{MDA}_3^{\star(j)},$$

where M is the number of trees in the forest. Additionally, if M tends to infinity, then

$$(iii) \quad \widehat{\text{MDA}}_{M,n}^{(IK)}(X^{(j)}) \xrightarrow{\mathbb{L}^1} \mathbb{V}[Y] \times ST^{(j)} + \text{MDA}_3^{\star(j)}.$$

In the sequel, we denote $\text{MDA}_1^{\star(j)} = \mathbb{V}[Y] \times ST^{(j)}$ and $\text{MDA}_2^{\star(j)} = \mathbb{V}[Y] \times ST_{mg}^{(j)}$. Each term of the decompositions of Proposition 3.1 can be interpreted alone.

$\text{MDA}_1^{\star(j)}$ is the non-normalized total Sobol index that has a straightforward interpretation: the amount of explained output variance lost when $X^{(j)}$ is removed from the model. This quantity is really the information one is looking for when computing the MDA for objective (i).

$\text{MDA}_2^{\star(j)}$ is the non-normalized marginal total Sobol index. Its interpretation is more difficult. Intuitively, in the case of $\text{MDA}_1^{\star(j)}$, contributions due to the dependence between $X^{(j)}$ and $\mathbf{X}^{(-j)}$ are excluded because of the conditioning on $\mathbf{X}^{(-j)}$. For $\text{MDA}_2^{\star(j)}$, this dependence is ignored, and therefore such removal does not take place. For example, if $X^{(j)}$ has a strong influence on the regression function but is highly correlated with other variables, then $\text{MDA}_1^{\star(j)}$ is small, whereas $\text{MDA}_2^{\star(j)}$ is high. For objective (i), one wants to keep only one variable of a group of highly influential and correlated inputs, and therefore $ST_{mg}^{(j)}$ can be a misleading component.

$\text{MDA}_3^{\star(j)}$ is not a known measure of importance, and seems to have no clear interpretation: it measures how the permutation shifts the average of m over the j-th input, and thus characterizes the structure of m and the dependence of $\mathbf{X}$ combined.

---

[4] See Bénard et al. [8] for details.

$\mathrm{MDA}_3^{*(j)}$ is null if variables are independent. The value of $\mathrm{MDA}_3^{*(j)}$ increases with dependence, and this effect can be amplified by interactions between variables.

Overall, all MDA definitions are misleading with respect to both objectives (i) selecting a small number of input variables with a maximized predictive accuracy and (ii) ranking all influential variables to focus on for further exploration with domain experts, since they include $\mathrm{MDA}_3^{*(j)}$ in their theoretical counterparts. For the first objective of variable importance, mentioned at the beginning of the section, we want to estimate the total Sobol index $\mathrm{MDA}_1^{*(j)}$, which gives the output variance lost when a given variable is removed from the model, and therefore enables to perform backward variable selection. Thus, we present the Sobol-MDA, a fast and consistent algorithm to estimate the total Sobol index, based on random forests.

3.3.2.1   Sobol-MDA Algorithm

The key feature of the original MDA procedures is to permute the values of the j-th component of the data to break its relation to the output, and then compute the degraded accuracy of the forest. Observe that this is strictly equivalent to drop the original dataset down each tree of the forest, but when a sample hits a split involving variable j, it is randomly sent to the left or right side with a probability equal to the proportion of points in each child node. This fact highlights that the goal of the MDA is simply to perturb the tree prediction process to cancel out the splits on variable j. Besides, notice that this point of view on the MDA procedure (using the original dataset and noisy trees) is introduced by Ishwaran [54] to conduct a theoretical analysis of a modified version of the MDA. Here, our Sobol-MDA algorithm builds on the same principle of ignoring splits on variable j, such that the noisy CART tree predicts $\mathbb{E}[m(\mathbf{X})|\mathbf{X}^{(-j)}]$ (instead of $m(\mathbf{X})$ for the original CART). It enables to recover the proper theoretical counterpart: the unnormalized total Sobol index, i.e., $\mathbb{E}[\mathbb{V}(m(\mathbf{X})|\mathbf{X}^{(-j)})]$. To achieve this, we leave aside the permutation trick, and use another approach to cancel out a given variable j in the tree prediction process: the partition of the input space obtained with the terminal leaves of the original tree is projected along the j-th direction—see Fig. 3.10, and the outputs of the cells of this new projected partition are recomputed with the training data. From an algorithmic point of view, this procedure is quite straightforward as we will see below, and enables to get rid of variable $X^{(j)}$ in the tree estimate. Then, it is possible to compute the accuracy of the associated Out-of-Bag (OOB) (see [12] for details) projected forest estimate, subtract it from the original accuracy, and normalize the obtained difference by $\mathbb{V}[Y]$ to obtain the Sobol-MDA for variable $X^{(j)}$.

Interestingly, to compute SHAP values for tree ensembles, Lundberg et al. [65] also introduce an algorithm to modify the CART predictions to estimate $\mathbb{E}[m(\mathbf{X})|\mathbf{X}^{(-j)}]$. More precisely, they propose the following recursive algorithm: the query point $\mathbf{x}$ is dropped down the tree, but when a split on variable j is hit, $\mathbf{x}$ is sent to both the left and right children nodes. Then, $\mathbf{x}$ falls in multiple terminal cells of the tree. The final prediction is the weighted average of the cell outputs, where the weight associated to a terminal leaf A is given by an estimate of

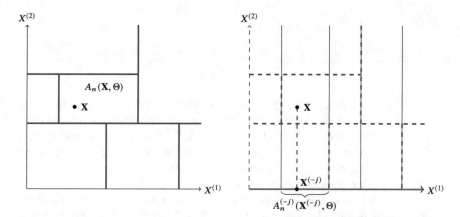

Fig. 3.10 Example of a partition produced by a random CART tree (left side) projected on the subspace span by $\mathbf{X}^{(-2)} = X^{(1)}$ (right side). Here, $p = 2$ and $j = 2$

$\mathbb{P}(\mathbf{X} \in A | \mathbf{X}^{(-j)} = \mathbf{x}^{(-j)})$: the product of the empirical probabilities to choose the side that leads to A at each split on variable j in the path of the original tree. At first sight, their approach seems suited to estimate total Sobol indices, but unfortunately, the weights are properly estimated by such procedure only if the components of $\mathbf{X}$ are independent. Therefore, as highlighted in Aas et al. [1], this algorithm gives biased predictions in a correlated setting.

We improve over Lundberg et al. [65] with the Projected-CART Algorithm: both training and out-of-bag samples are dropped down the tree and sent on both right and left children nodes when a split on variable j is met. Again, each data point may belong to multiple cells at each level of the tree. For each out-of-bag sample, the associated prediction is the output average over all training samples that belong to the same collection of terminal leaves. In other words, we compute the intersection of these terminal leaves to select the training observations belonging to every cell of this collection to estimate the prediction. This intersection gives the projected cell. This mechanism is equivalent to projecting the tree partition on the subspace span by $\mathbf{X}^{(-j)}$, as illustrated in Fig. 3.10 for $p = 2$ and $j = 2$. Recall that $A_n(\mathbf{X}, \Theta)$ is the cell of the original tree partition where $\mathbf{X}$ falls, whereas the associated cell of the projected partition is denoted $A_n^{(-j)}(\mathbf{X}^{(-j)}, \Theta)$. Formally, we respectively denote the associated projected tree and projected out-of-bag forest estimates as $m_n^{(-j)}(\mathbf{X}^{(-j)}, \Theta)$ and $m_{M,n}^{(-j,OOB)}(\mathbf{X}_i^{(-j)}, \Theta_M)$, respectively defined by

$$m_n^{(-j)}(\mathbf{X}^{(-j)}, \Theta) = \frac{\sum_{i=1}^{a_n} Y_i \mathbb{1}_{\mathbf{X}_i \in A_n^{(-j)}(\mathbf{X}^{(-j)}, \Theta)}}{\sum_{i=1}^{a_n} \mathbb{1}_{\mathbf{X}_i \in A_n^{(-j)}(\mathbf{X}^{(-j)}, \Theta)}},$$

and for $i \in \{1, \ldots, n\}$,

$$m_{M,n}^{(-j,OOB)}(\mathbf{X}_i^{(-j)}, \boldsymbol{\Theta}_M) = \frac{1}{|\Lambda_{n,i}|} \sum_{\ell \in \Lambda_{n,i}} m_n^{(-j)}(\mathbf{X}_i^{(-j)}, \Theta_\ell) \mathbb{1}_{|\Lambda_{n,i}|>0}.$$

The Projected-CART algorithm provides two sources of improvements over Lundberg et al. [65]: first, the training data points are dropped down the modified tree to recompute the cell outputs, and thus $\mathbb{E}[m(\mathbf{X})|\mathbf{X}^{(-j)} \in A]$ is directly estimated in each cell. Secondly, the projected partition is finer than in the original tree, which mitigates masking effects (when an influential variable is not often selected in the tree splits because of other highly correlated variables).

Finally, the Sobol-MDA estimate is given by the normalized difference of the quadratic error of the OOB projected forest with the OOB error of the original forest. Formally, we define the Sobol-MDA as

$$S\widehat{-MDA}_{M,n}(X^{(j)}) = \frac{1}{\hat{\sigma}_Y^2} \frac{1}{n} \sum_{i=1}^{n}(Y_i - m_{M,n}^{(-j,OOB)}(\mathbf{X}_i^{(-j)}, \boldsymbol{\Theta}_M))^2 - (Y_i - m_{M,n}^{(OOB)}(\mathbf{X}_i, \boldsymbol{\Theta}_M))^2,$$

where $\hat{\sigma}_Y^2 = \frac{1}{n-1}\sum_{i=1}^{n}(Y_i - \bar{Y})^2$ is the standard variance estimate of the output Y. An implementation in R and C++ of the Sobol-MDA is available at https://gitlab.com/drti/sobolmda and is based on `ranger` [99].

3.3.2.2 Sobol-MDA Properties

Computational Complexity By definition, an estimate of the total Sobol index is given by the following procedure: retrain the random forest without the j-th variable, and subtract the associated explained variance to the original accuracy with all variables. However, this brute force approach is computationally expensive since it requires to fit p forests to get the total Sobol index of each variable. Louppe [63] states that the average computational complexity of the forest growing is $O(Mpn\log^2(n))$. Thus, the total complexity of the brute force approach is $O(Mp^2n\log^2(n))$, which is quadratic with the dimension p and therefore intractable in high-dimensional settings.

On the other hand, the original MDA procedure has an average complexity of $O(Mpn\log(n))$: to run a balanced tree prediction for a given data point, it is dropped down the $\log(n)$ levels of the tree, which makes a complexity of $O(n\log(n))$ for the full OOB sample, repeated for the M trees of the forest and the p variables. In the Sobol-MDA procedure, the complexity analysis is similar, except that when a point is dropped down the tree, it can be sent to both the left and right children nodes, generating multiple operations at a given tree level and then an additional multiplicative factor of $\log(n)$. However, it is not necessary to run the Projected-CART algorithm for each of the p variables. Indeed, when a given observation is dropped down the tree, it meets at most $\log(n)$ different variables in the original tree path. Therefore, the Projected-CART prediction has to be computed only for $\log(n)$ variables for each observation. Thus, the Sobol-MDA algorithm has a computational complexity of $O(Mn \log^3(n))$, which is in particular independent of the dimension p, and quasi-linear with the sample size n.

Consistency The original MDA versions do not converge towards the total Sobol index, which is the relevant quantity for our objective (i)—see Proposition 3.1. On the other hand, the Sobol-MDA is consistent as stated in Theorem 2.2 in [8]. This is the first result to establish the convergence of the empirical MDA, computed with random forest algorithm (up to minor modifications), toward a theoretical quantity, useful to rank influential variables.

3.3.2.3 Experiments

First, we consider an analytical example on which the true total Sobol indices can be computed. In this setting, we show empirically that the Sobol-MDA leads to the accurate importance variable ranking, while original MDA versions do not. Next, we apply the RFE on real data to show the performance improvement of the Sobol-MDA for variable selection.

Simulated Data We consider here the following example, where the data has both dependence and interactions. In our example, the input is a Gaussian vector of zero mean and with $p = 5$, and the regression function is given by

$$m(\mathbf{X}) = \alpha X^{(1)} X^{(2)} \mathbb{1}_{X^{(3)}>0} + \beta X^{(4)} X^{(5)} \mathbb{1}_{X^{(3)}<0}.$$

Here, we set $\alpha = 1.5$, $\beta = 1$, $\mathbb{V}[X^{(j)}] = 1$ for all variables $j \in \{1, \ldots, 5\}$, and the correlation coefficients are set to $\rho_{1,2} = 0.9$ and $\rho_{4,5} = 0.6$ (other covariance terms are null). Finally, we define the model output as $Y = m(\mathbf{X}) + \varepsilon$, where ε is an independent centered gaussian noise whose variance verifies $\mathbb{V}[\varepsilon]/\mathbb{V}[Y] = 10\%$. Then, we run the following experiment: first, we generate a sample $\mathcal{D}_n$ of size $n = 3000$ and distributed as the Gaussian vector $\mathbf{X}$. Next, a random forest of $M = 300$ trees is fit with $\mathcal{D}_n$ and we compute the BC-MDA, IK-MDA, and Sobol-MDA. To enable comparisons, the BC-MDA is normalized by $2\mathbb{V}[Y]$, and the IK-MDA by $\mathbb{V}[Y]$—see Proposition 3.1. To show the improvement of our Projected-CART algorithm, we also compute the Sobol-MDA using the algorithm from Lundberg et al. [65], denoted $S - \widehat{MDA}_{Ldg}$. All results are reported in Table 3.3 and the theoretical counterparts of the estimates are also provided. Notice that the associated standard deviations are gathered in Table 3.4, and that the variables are ranked by decreasing values of the theoretical total Sobol index since it is the value of interest: $\mathbf{X}^{(3)}$, then $\mathbf{X}^{(4)}$ and $\mathbf{X}^{(5)}$, and finally $\mathbf{X}^{(1)}$ and $\mathbf{X}^{(2)}$.

Only the Sobol-MDA computed with the Projected-CART algorithm ranks the variables in the same appropriate order than the total Sobol index. In particular,

Table 3.3 Normalized BC-MDA, normalized IK-MDA, and Sobol-MDA estimates for Example 1

	BC − MDA*	$\widehat{BC - MDA}$	IK − MDA*	$\widehat{IK - MDA}$	ST*	$\widehat{S - MDA}$	$\widehat{S - MDA}_{Ldg}$
$X^{(3)}$	0.47	0.37	0.47	0.43	0.47	0.45	0.43
$X^{(4)}$	0.21	0.10	0.37	0.14	0.10	0.08	0.13
$X^{(5)}$	0.21	0.09	0.37	0.13	0.10	0.08	0.13
$X^{(1)}$	0.64	0.24	1.0	0.29	0.07	0.05	0.22
$X^{(2)}$	0.64	0.24	1.0	0.28	0.07	0.05	0.23

Table 3.4 Standard deviations of the normalized BC-MDA, normalized IK-MDA, and Sobol-MDA estimates over 10 repetitions for Example 1

	$\widehat{IK - MDA}$	$\widehat{BC - MDA}$	$\widehat{S - MDA}$	$\widehat{S - MDA}_{Ldg}$
$X^{(3)}$	0.02	0.03	0.03	0.03
$X^{(4)}$	0.01	0.02	0.01	0.01
$X^{(5)}$	0.01	0.01	0.01	0.01
$X^{(1)}$	0.02	0.02	0.01	0.02
$X^{(2)}$	0.02	0.02	0.01	0.01

$X^{(4)}$ and $X^{(5)}$ have a higher total Sobol index than variables 1 and 2 because of the stronger correlation between $X^{(1)}$ and $X^{(2)}$ than between $X^{(4)}$ and $X^{(5)}$. For all the other importance measures, $X^{(1)}$ and $X^{(2)}$ are more important than $X^{(4)}$ and $X^{(5)}$. For the original MDA, this is due to the higher coefficient $\alpha = 1.5 > \beta = 1$, to the term $MDA_2^{\star(j)}$, and especially to $MDA_3^{\star(j)}$ which increases with correlation. Since the explained variance of the random forest is 82% in this experiment, all estimates have a negative bias. The bias of the BC-MDA and IK-MDA dramatically increases with correlation. Indeed, a strong correlation between variables leaves some regions of the input space free of training data. However, the OOB (Out-Of-Bag) permuted sample queries the forest in these regions where the forest extrapolates. This phenomenon combined with the $MDA_3^{\star(j)}$ component explains the high bias of the BC-MDA and IK-MDA for correlated inputs. Also observe that since $X^{(3)}$ is independent of the other variables, the bias is small for both the BC/IK-MDA, and it is smaller for the IK-MDA than the BC-MDA as the forest estimate is more accurate than a single tree. Finally, the Sobol-MDA computed with the algorithm in [65] is biased as suggested in [1], and the bias also seems to increase with correlation.

Recursive Feature Elimination The Recursive Feature Elimination algorithm (RFE) is originally introduced by Guyon et al. [49] to perform variable selection with SVM. Gregorutti et al. [47] apply RFE to random forests with the MDA as importance measure. The principle of the RFE algorithm is to discard the less relevant input variables one by one, based on their MDA values, which are recomputed at each step. Thus, the RFE is a relevant strategy for our objective (i) of building a model with a high accuracy and a small number of variables. At each step of the RFE, the goal is to detect the less relevant input variable based on the trained model. Since the total Sobol index measures the proportion of explained output variance lost when a given variable is removed, the optimal strategy is therefore to discard the variable with the smallest total Sobol index. As the Sobol-MDA directly estimates the total Sobol index whereas existing MDA all have additional noisy terms (see Proposition 3.1), using the Sobol-MDA improves the performance of the RFE, as shown in the following experiments.

The RFE algorithm is illustrated with two real datasets: following [45] we use the "Ozone" data [32] for a regression example, as well as the "Breast Cancer Wisconsin Diagnosis" data for a binary classification case as in Song et al. [89]. The RFE is run three times, respectively using the BC-MDA, IK-MDA, and the Sobol-MDA as importance measures to iteratively discard the less relevant variable. At each step of the RFE, the explained variance of the forest is retrieved. As explained in Gregorutti et al. [47], using the OOB error gives optimistically bias results in this context, since the same data is used to compute both the forest error and the importance measures. Instead, we use a 10-fold cross-validation: the forest and the associated importance measure are computed with 9 folds, and the error is estimated with the 10-th fold. For each dataset, the cross-validation is repeated 40 times to get the result uncertainties, displayed as boxplots in the figures. Figure 3.11 highlights that the Sobol-MDA leads to a more efficient variable selection than the BC-MDA and the IK-MDA for the "Ozone" and "Breast Cancer Wisconsin Diagnosis" datasets. Notice that the IK-MDA performs better than the BC-MDA, as expected from their theoretical counterparts—see Proposition 3.1.

3.3.3   SHAFF: SHApley eFfects Estimates via Random Forests

We have identified two possible final aims when computing variable importance: (i) reducing the problem dimension by selecting a small number of input variables with a maximized predictive accuracy or (ii) detecting and ranking all influential variables to focus on for further exploration with domain experts. Sobol-MDA improves upon the standard MDA variable importance measure in random forests if one aims at estimating the Sobol indices, which are the theoretical quantities to use to fulfill our first objective (i).

To achieve the second objective (ii), Sobol indices are not the correct quantities of interest since they do not take into account the mutual information between one given variable and the other ones. Shapley effects are now widely used to handle objective (ii), as they can efficiently handle dependence and interactions in the data, as opposed to most other variable importance measures. However, estimating Shapley effects is a challenging task, because of the computational complexity and the conditional expectation estimates. In this section, we leverage random forests to develop SHAFF, a fast and accurate Shapley effect estimate. Such remarkable performance is reached by combining two new features. Firstly, we improve the Monte-Carlo approach by using importance sampling to focus

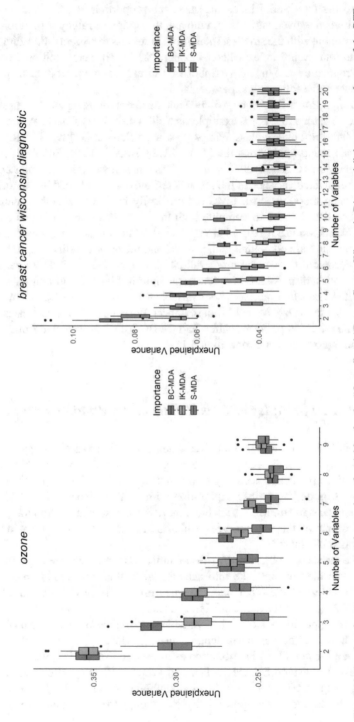

Fig. 3.11 Random forest error versus the number of variables for the "Ozone" and "Breast Cancer Wisconsin Diagnostic" datasets at each step of the RFE, using different importance measures: BC-MDA, IK-MDA, and Sobol-MDA (Figure 3 in https://arxiv.org/pdf/2102.13347v1.pdf)

on the most relevant subsets of variables identified by the forest. Secondly, we extend our projected random forest algorithm to compute fast and accurate estimates of the conditional expectations for any variable subset. The algorithm details are provided in Sect. 3.3.3.1. Next, we prove the consistency of SHAFF in Sect. 3.3.3.2. To our knowledge, SHAFF is the first Shapley effect estimate, which is both computationally fast and consistent in a general setting. In Sect. 3.3.3.3, several experiments show the practical improvement of our method over state-of-the-art algorithms. Additional information on SHAFF can be found in [6].

3.3.3.1 SHAFF Algorithm

Existing Approach SHAFF builds on two Shapley algorithms: Lundberg and Lee [64], Williamson and Feng [98]. From these approaches, we can deduce the following general three-step procedure to estimate Shapley effects. First, a set $\mathcal{U}_{n,K}$ of K variable subsets $U \subset \{1, \ldots, p\}$ is randomly drawn. Next, an estimate $\hat{v}_n(U)$ of $\mathbb{V}[\mathbb{E}[Y|\mathbf{X}^{(U)}]]$ is computed for all selected U from an available sample $\mathcal{D}_n = \{(\mathbf{X}_1, Y_1), \ldots, (\mathbf{X}_n, Y_n)\}$ of n independent random variables distributed as $(\mathbf{X}, Y)$. Finally, Shapley effects are defined as the least square solution of a weighted linear regression problem. If $I(U)$ is the binary vector of dimension p where the j-th component takes the value 1 if $j \in U$ and 0 otherwise, Shapley effect estimates are the minimum in β of the following cost function:

$$\ell_n(\beta) = \frac{1}{K} \sum_{U \in \mathcal{U}_{n,K}} w(U)(\hat{v}_n(U) - \beta^T I(U))^2,$$

where the weights $w(U)$ are given by

$$w(U) = \frac{p-1}{\binom{p}{|U|}|U|(p-|U|)},$$

and the coefficient vector β is constrained to have its components sum to $\hat{v}_n(\{1, \ldots, p\})$.

Algorithm Overview SHAFF introduces two new critical features to estimate Shapley effects efficiently, using an initial random forest model. Firstly, we apply importance sampling to select variable subsets $U \subset \{1, \ldots, p\}$, based on the variables frequently selected in the forest splits. This favors the sampling of subsets U containing influential and interacting variables. Secondly, for each selected subset U, the variance of the conditional expectation is estimated with the projected forest algorithm described below, which is both a fast and consistent approach. We will see that these features considerably reduce the computational cost and the estimate error. To summarize, once an initial random forest is fit, SHAFF proceeds in three steps:

1. sample many subsets U, typically a few hundreds, based on their occurrence frequency in the random forest;
2. estimate $\mathbb{V}[\mathbb{E}[Y|\mathbf{X}^{(U)}]]$ with the projected forest algorithm for all selected U and their complementary sets $\{1, \ldots, p\} \setminus U$;
3. solve a weighted linear regression problem to recover Shapley effects.

Initial Random Forest Prior to SHAFF, a random forest is fit with the training sample $\mathscr{D}_n$ to generalize the relation between the inputs $\mathbf{X}$ and the output Y. A large number M of CART trees are averaged to form the final forest estimate $m_{M,n}(\mathbf{x}, \boldsymbol{\Theta}_M)$, where $\mathbf{x}$ is a new query point, and each tree is randomized by a component of $\boldsymbol{\Theta}_M = (\Theta_1, \ldots, \Theta_\ell, \ldots, \Theta_M)$. Each Θ_ℓ is used to bootstrap the data prior to the ℓ-th tree growing, and to randomly select mtry variables to optimize the split at each node. mtry is a parameter of the forest, and its efficient default value is $p/3$. In the sequel, we will need the forest parameter min_node_size, which is the minimum number of observations in a terminal cell of a tree, as well as the out-of-bag (OOB) sample of the ℓ-th tree: the observations which are left aside in the bootstrap sampling prior to the construction of tree ℓ. Given this initial random forest, we can now detail the main three steps of SHAFF.

Importance Sampling The Shapley effect formula for a given variable $X^{(j)}$ sums terms over all subsets of variables $U \subset \{1, \ldots, p\} \setminus \{j\}$, which makes 2^{p-1} terms, an intractable problem in most cases. SHAFF uses importance sampling to draw a reasonable number of subsets U, typically a few hundreds, while preserving a high accuracy of the Shapley estimates. We take advantage of the initial random forest to define an importance measure for each variable subset U, used as weights for the importance sampling distribution.

In a tree construction, the best split is selected at each node among mtry input variables. Therefore, as highlighted by Proposition 1 in [82], the forest naturally splits on influential variables. SHAFF leverages this idea to define an importance measure for all variable subsets $U \subset \{1, \ldots, p\}$ as the probability that a given U occurs in a path of a tree of the forest. Empirically, this means that we count the occurrence frequency of U in the paths of the M trees of the forest, and denote it by $\hat{p}_{M,n}(U)$. Such approach is inspired by Basu et al. [4] and Bénard et al. [7]. This principle is illustrated with the following simple example in dimension $p = 10$. Let us consider a tree, where the root node splits on variable $X^{(5)}$, the left child node splits on variable $X^{(3)}$, and the subsequent left child node at the third tree level, on variable $X^{(2)}$. Thus, the path that leads to the extreme left node at the fourth level uses the following index sequence of splitting variables: $\{5, 3, 2\}$. All in all, the following variable subsets are included in this tree path: $U = \{5\}$, $U = \{3, 5\}$, and $U = \{2, 3, 5\}$. Then, SHAFF runs through the forest to count the number of times each subset U occurs in the forest paths, and computes the associated frequency $\hat{p}_{M,n}(U)$. If a subset U does not occur in the forest, we obviously have $\hat{p}_{M,n}(U) = 0$. Notice that the computational complexity of this step is linear: $O(Mn)$.

The occurrence frequencies $\hat{p}_{M,n}(U)$ defined above are scaled to sum to 1, and then define a discrete distribution for the set of all subsets of variables $U \subset \{1, \ldots, p\}$, excluding the full and empty sets. By construction, this distribution is skewed towards the subsets U containing influential variables and interactions, and is used for the importance sampling. Finally, SHAFF draws a number K of subsets U with respect to this discrete distribution, where K is a hyperparameter of the algorithm. We define $\mathcal{U}_{n,K}$ the random set of the selected variable subsets U. For all $U \in \mathcal{U}_{n,K}$, SHAFF also includes the complementary set $\{1, \ldots, p\} \setminus U$ in $\mathcal{U}_{n,K}$, as Covert and Lee [25] show that this "paired sampling" improves the final Shapley estimate accuracy. Clearly, the computational complexity and the accuracy of the algorithm increase with K. The next step of SHAFF is to efficiently estimate $\mathbb{V}[\mathbb{E}[Y|\mathbf{X}^{(U)}]]$ for all drawn $U \in \mathcal{U}_{n,K}$.

Projected Random Forests In order to estimate $\mathbb{V}[\mathbb{E}[Y|\mathbf{X}^{(U)}]]$ for the selected variable subsets $U \in \mathcal{U}_{n,K}$, most existing methods use greedy algorithms. However, such estimates are not accurate in moderate or large dimensions when input variables are dependent [1, 92]. Another approach is to train a new model for each subset U, but this is computationally costly [98]. To solve this issue, we use our projected random forest algorithm (PRF), introduced in the previous section, and generalize it in order to obtain a fast and accurate estimate of $\mathbb{V}[\mathbb{E}[Y|\mathbf{X}^{(U)}]]/\mathbb{V}[Y]$ for any variable subset $U \subset \{1, \ldots, p\}$.

The computational complexity of PRF for all $U \in \mathcal{U}_{n,K}$ does not depend on the dimension p, is linear with M, K, and quasi-linear with n: $O(MKn\log(n))$. PRF is therefore faster than growing K random forests from scratch, one for each subset U, which has an averaged complexity of $O(MKpn\log^2(n))$ [63]. The computational gain of SHAFF can be considerable in high dimension, since the complexity of all competitors depends on p. Notice that the PRF algorithm is close in spirit to a component of the Sobol-MDA [8], used to measure the loss of output explained variance when an input variable j is removed from a random forest. In particular, a naive adaptation leads to a quadratic complexity with respect to the sample size n, whereas our PRF algorithm has a quasi-linear complexity, which makes it operational. Finally, the last step of SHAFF is to take advantage of the estimated $\hat{v}_{M,n}(U)$ for $U \in \mathcal{U}_{n,K}$ to recover Shapley effects.

Shapley Effect Estimates The importance sampling introduces the corrective terms $\hat{p}_{M,n}(U)$ in the final loss function. Thus, SHAFF estimates $\widehat{\mathbf{Sh}}_{M,n} = (\hat{Sh}_{M,n}(X^{(1)}), \ldots, \hat{Sh}_{M,n}(X^{(p)}))$ as the minimum in β of the following cost function

$$\ell_{M,n}(\beta) = \frac{1}{K} \sum_{U \in \mathcal{U}_{n,K}} \frac{w(U)}{\hat{p}_{M,n}(U)} (\hat{v}_{M,n}(U) - \beta^T I(U))^2,$$

where the sum of the components of β is constrained to be the proportion of output explained variance of the initial forest, fit with all input variables. Finally, this can be written in the following compact form:

$$\hat{\mathbf{Sh}}_{M,n} = \underset{\beta \in [0,1]^p}{\text{argmin}} \quad \ell_{M,n}(\beta)$$

$$\text{s.t.} ||\beta||_1 = \hat{v}_{M,n}(\{1, \ldots, p\}).$$

3.3.3.2 SHAFF Consistency

We prove in this section that SHAFF is consistent, in the sense that the estimated value is arbitrarily close to the ground truth theoretical Shapley effect, provided that the sample size is large enough. To our knowledge, we provide the first Shapley algorithm which requires to fit only a single initial model and is consistent in the general case. We insist that our result is valid even when input variables exhibit strong dependences. The consistency of SHAFF holds under the following mild and standard assumption on the data distribution:

(A3) The response $Y \in \mathbb{R}$ follows

$$Y = m(\mathbf{X}) + \varepsilon,$$

where $\mathbf{X} = (X^{(1)}, \ldots, X^{(p)}) \in [0, 1]^p$ admits a density over $[0, 1]^p$ bounded from above and below by strictly positive constants, m is continuous, and the noise ε is sub-Gaussian, independent of $\mathbf{X}$, and centered. □

To alleviate the mathematical analysis, we slightly modify the standard Breiman random forests: the bootstrap sampling is replaced by a subsampling without replacement of a_n observations, as it is usually done in the theoretical analysis of random forests [69, 82]. Additionally, we follow Wager and Athey [96] with an additional small modification of the forest algorithm, which is sufficient to ensure its consistency. Firstly, a node split is constrained to generate child nodes with at least a small fraction $\gamma > 0$ of the parent node observations. Secondly, the split selection is slightly randomized: at each tree node, the number mtry of candidate variables drawn to optimize the split is set to mtry $= 1$ with a small probability $\delta > 0$. Otherwise, with probability $1 - \delta$, the default value of mtry is used. It is stressed that these last modifications are mild, since γ and δ can be chosen arbitrarily small.

Finally, we introduce the following two assumptions on the asymptotic regime of the algorithm parameters. Assumption (A3.3.3.2) enforces that the tree partitions are not too complex with respect to the sample size n. On the other hand, Assumption (A3.3.3.2) states that the number of trees and the number of sampled variable subsets U grow with n. This ensures that all possible variable subsets have a positive probability to be drawn, which is required for the convergence of our algorithm based on importance sampling.

(A4) The asymptotic regime of a_n, the size of the subsampling without replacement, and the number of terminal leaves t_n are such that $a_n \leq n - 2$, $a_n/n < 1 - \kappa$ for a fixed $\kappa > 0$, $\lim_{n \to \infty} a_n = \infty$, $\lim_{n \to \infty} t_n = \infty$, and $\lim_{n \to \infty} 2^{t_n} \frac{(\log(a_n))^9}{a_n} = 0$. □

(A5) The number of Monte-Carlo sampling K_n and the number of trees M_n grow with n, such that $M_n \longrightarrow \infty$ and $n.M_n/K_n \longrightarrow 0$. □

We also let the theoretical Shapley effect vector be $\mathbf{Sh}^\star = (Sh^{(1)}, \dots, Sh^{(p)})$. According to Theorem 3.2 below, the empirical Shapley effects, computed via the procedure described above, is consistent.

Theorem 3.2 ([6]) *If Assumptions (A3.3.3.2), (A3.3.3.2), and (A3.3.3.2) are satisfied, then SHAFF is consistent, that is* $\hat{\mathbf{Sh}}_{M_n,n} \xrightarrow{P} \mathbf{Sh}^\star$.

3.3.3.3 Experiments

We run two batches of experiments to show the improvements of SHAFF over the main competitors Broto et al. [17], Williamson and Feng [98], and Covert et al. [26]. Experiment 1 is a simple linear case with a redundant variable, while Experiment 2 is a non-linear example with high-order interactions. In both cases, existing Shapley algorithms exhibit a bias which significantly modifies the accurate variable ranking, as opposed to SHAFF. Finally, we combine the new features of SHAFF with existing algorithms to break down the performance improvements due to the importance sampling and the projected forest.

Experiment Settings Our implementation of SHAFF is based on `ranger`, a fast random forest software written in C++ and R from Wright and Ziegler [99]. We implemented Williamson and Feng [98] from scratch, as it only requires to sample variable subsets U, fit a random forest for each U, and recover Shapley effects by solving the linear regression problem defined in Sect. 3.3.3.1. Notice that we limit tree depth to 6 when $|U| \leq 2$ to avoid overfitting. We implemented SAGE following Algorithm 1 from Covert et al. [26], and setting $m = 30$. The original implementation of Broto et al. [17] in the R package `sensitivity` has an exponential complexity with p. Even for $p = 10$, we could not have the experiments done within 24 hours when parallelized on 16 cores. Therefore, we do not display the results for Broto et al. [17], which seem to have a high bias on toy examples. In all procedures, the number K of sampled subsets U is set to 500, and we use 500 trees for the forest growing. Each run is repeated 30 times to estimate the standard deviations. For both experiments, we analytically derive the theoretical Shapley effects, and display this ground truth with red crosses in Fig. 3.12.

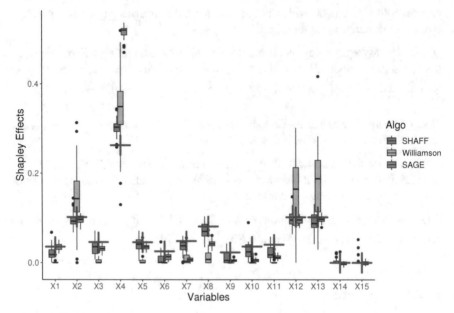

Fig. 3.12 Shapley effects for Experiment 1. Red crosses are the theoretical Shapley effects (Figure 2 in https://arxiv.org/pdf/2105.11724v3.pdf)

Experiment 1 In the first experiment, we consider a linear model and a correlated centered Gaussian input vector of dimension 11. The output Y follows

$$Y = \beta^T \mathbf{X} + \varepsilon,$$

where $\beta \in [0, 1]^{11}$, and the noise ε is centered, independent, and such that $\mathbb{V}[\varepsilon] = 0.05 \times \mathbb{V}[Y]$. Finally, two copies of $X^{(2)}$ are appended to the data as $X^{(12)}$ and $X^{(13)}$, and two dummy Gaussian variables $X^{(14)}$ and $X^{(15)}$ are also added. We draw a sample $\mathscr{D}_n$ of size $n = 3000$.

Figure 3.12 shows that SHAFF is more accurate than its competitors. Covert et al. [26] has a strong bias for several variables, in particular $X^{(4)}$, $X^{(7)}$, $X^{(8)}$, and $X^{(10)}$. The algorithm from Williamson and Feng [98] has a lower performance since its variance is higher than for the other methods. Notice that Williamson and Feng [98] recommend to set $K = 2n$ (= 6000 here). Since we use $K = 500$ to compare all algorithms, this high variance is quite expected and show the improvement due to the importance sampling of our method. Besides, the computational complexity of Williamson and Feng [98] is $O(n^2)$ whereas SHAFF is quasi-linear. Finally, in this experiment, the random forest has a proportion of explained variance of about 86%, and the noise variance is 5%, which explains the small negative bias of many estimated values.

Table 3.5 Cumulative absolute error of SHAFF versus state-of-the-art Shapley algorithms

Algorithm	Experiment 1	Experiment 2
SHAFF	0.25	0.15
Williamson	0.64	0.63
SAGE	0.33	0.18

Experiment 2 In the second experiment, we consider two independent blocks of 5 interacting variables. The input vector is Gaussian, centered, and of dimension 10. All variables have unit variance, and all covariances are null, except $\mathrm{Cov}(X^{(1)}, X^{(2)}) = \mathrm{Cov}(X^{(6)}, X^{(7)}) = 0.9$, and $\mathrm{Cov}(X^{(4)}, X^{(5)}) = \mathrm{Cov}(X^{(9)}, X^{(10)}) = 0.5$. The output Y follows

$$Y = 3\sqrt{3} \times X^{(1)} X^{(2)} \mathbb{1}_{X^{(3)}>0} + \sqrt{3} \times X^{(4)} X^{(5)} \mathbb{1}_{X^{(3)}<0}$$
$$+ 3 \times X^{(6)} X^{(7)} \mathbb{1}_{X^{(8)}>0} + X^{(9)} X^{(10)} \mathbb{1}_{X^{(8)}<0} + \varepsilon,$$

where the noise ε is centered, independent, and such that $\mathbb{V}[\varepsilon] = 0.05 \times \mathbb{V}[Y]$. We add 5 dummy Gaussian variables $X^{(11)}$, $X^{(12)}$, $X^{(13)}$, $X^{(14)}$, and $X^{(15)}$, and draw a sample $\mathscr{D}_n$ of size $n = 10000$.

In this context of strong interactions and correlations, we observe in Table 3.5 that SHAFF outperforms its competitors. Furthermore, SHAFF properly identifies variable $X^{(3)}$ as the most important one, whereas SAGE considerably overestimates the Shapley effects of variables $X^{(1)}$ and $X^{(2)}$.

SHAFF Analysis Table 3.6 displays the cumulative absolute error of Shapley algorithms, based on various combinations of variable subset sampling and conditional expectation estimates, for Experiments 1 and 2. The goal is to break down the improvement of SHAFF between the new features proposed in Sect. 3.3.3.1. Firstly, we compare two approaches for the variable subset sampling: our paired importance sampling procedure (pIS) introduced in Sect. 3.3.3.1, and the paired Monte-Carlo sampling (pMC) approach of Covert and Lee [25]. Secondly, we compare several estimates of the conditional expectations: our projected random forest introduced in Sect. 3.3.3.1, the brute force retraining of a random forest for each subset U (Forest) as in Williamson and Feng [98], the marginal sampling (Marginals) used in the SAGE approach [26], and the approach from Lundberg et al. [65] specific to tree ensembles (TreeSHAP). In all cases, Shapley estimates are recovered using step 3 defined in Sect. 3.3.3.1. The comparisons of the first and last two lines of Table 3.6 clearly show the large improvement due to the importance sampling of SHAFF, since the cumulative error is divided by two compared to the paired Monte-Carlo sampling and using identical conditional expectation estimates. We also observe that the PRF algorithm is competitive with the brute force method of retraining many random forests, with a much smaller computational cost. Additionally, although the TreeSHAP algorithm [65] is fast, it comes at the price of a much stronger bias than the other approaches. Finally, the marginal sampling is as efficient as PRF

Table 3.6 Cumulative absolute error of Shapley estimates, based on various strategies for variable subset sampling and conditional expectation estimates

Algorithm	Experiment 1	Experiment 2
SHAFF	0.25	0.15
pIS/Forest	0.23	0.23
pIS/Marginals	0.26	0.31
pIS/TreeSHAP	1.18	1.49
pMC/Projected-RF	0.55	0.29
pMC/Forest	0.56	0.50

for Experiment 1 where the regression function is linear, but it is not the case for Experiment 2 where variables have interactions.

3.3.4 Discussion

As highlighted by Genuer et al. [45], two objectives can be identified when computing variable importances: (i) selecting a small number of input variables with a maximized predictive accuracy or (ii) ranking all influential variables to focus on for further exploration with domain experts.

With respect to the first objective, Sobol-MDA improves upon the MDA versions computed in random forests: Sobol-MDA is both consistent and outperform empirically its competitors. From a theoretical perspective, a major hypothesis of the analysis is to assume the consistency of the theoretical trees of the random forest. Indeed, it is rather natural to consider a consistent estimate in order to analyze the variable importance it produces. However, work remains to be done to prove that this consistency condition is in fact necessary or if there could exist models in which the variable importances are consistent and tree estimates are not.

With respect to the second objective (ii), we have proved that SHAFF is an efficient algorithm to estimate Shapley effects, that can be used to this aim. However, Shapley effects may not be the correct quantity to estimate in all settings [57]. Tailoring the proposed estimation of Shapley effects to specific settings, in particular that of causal inference, is definitely an interesting area of research.

Acknowledgments We would like to thank the many referees that helped us to improve the overall quality of the papers on which this chapter is built. We also want to express our warm thanks to Gerard Biau for his work and his numerous ideas in the presented work.

References

1. Aas K, Jullum M, Løland A (2019) Explaining individual predictions when features are dependent: More accurate approximations to Shapley values. Preprint. arXiv:190310464
2. Alelyani S, Zhao Z, Liu H (2011) A dilemma in assessing stability of feature selection algorithms. In: 13th IEEE international conference on high performance computing & communication. IEEE, Piscataway, pp 701–707
3. Archer K, Kimes R (2008) Empirical characterization of random forest variable importance measures. Comput Stat Data Anal 52:2249–2260
4. Basu S, Kumbier K, Brown J, Yu B (2018) Iterative random forests to discover predictive and stable high-order interactions. Proc Natl Acad Sci 115:1943–1948
5. Bénard C, Biau G, Da Veiga S, Scornet E (2021) Interpretable random forests via rule extraction. In: International Conference on Artif Intell Stat PMLR:937–945
6. Bénard C, Biau G, Da Veiga S, Scornet E (2021) SHAFF: Fast and consistent SHApley eFfect estimates via random Forests. Preprint. arXiv:210511724
7. Bénard C, Biau G, Da Veiga S, Scornet E (2021) SIRUS: Stable and Interpretable RUle Set for classification. Electron J Stat 15:427–505
8. Bénard C, Da Veiga S, Scornet E (2021) MDA for random forests: inconsistency, and a practical solution via the Sobol-MDA. Preprint. arXiv:210213347
9. Boulesteix AL, Slawski M (2009) Stability and aggregation of ranked gene lists. Brief Bioinform 10:556–568
10. Bousquet O, Elisseeff A (2002) Stability and generalization. J Mach Learn Res 2:499–526
11. Breiman L (1996) Bagging predictors. Mach Learn 24:123–140
12. Breiman L (1996) Out-of-bag estimation. Technical report, Statistics Department, University of California Berkeley
13. Breiman L (2001) Random forests. Mach Learn 45:5–32
14. Breiman L (2001) Statistical modeling: the two cultures (with comments and a rejoinder by the author). Stat Sci 16:199–231
15. Breiman L (2003) Setting up, using, and understanding random forests v3.1. https://www.stat.berkeley.edu/~breiman/Using_random_forests_V3.1.pdf
16. Breiman L, Friedman J, Olshen R, Stone C (1984) Classification and regression trees. Chapman & Hall/CRC, Boca Raton
17. Broto B, Bachoc F, Depecker M (2020) Variance reduction for estimation of Shapley effects and adaptation to unknown input distribution. SIAM/ASA J Uncertain Quant 8:693–716
18. Candes E, Fan Y, Janson L, Lv J (2016) Panning for gold: Model-X knockoffs for high-dimensional controlled variable selection. Preprint. arXiv:161002351
19. Chao A, Chazdon R, Colwell R, Shen TJ (2006) Abundance-based similarity indices and their estimation when there are unseen species in samples. Biometrics 62:361–371
20. Chastaing G, Gamboa F, Prieur C (2012) Generalized Hoeffding-Sobol decomposition for dependent variables-application to sensitivity analysis. Electron J Stat 6:2420–2448
21. Chen T, Guestrin C (2016) Xgboost: a scalable tree boosting system. In: Proceedings of the 22nd ACM SIGKDD international conference on knowledge discovery and data mining. ACM, New York, pp 785–794
22. Clark P, Niblett T (1989) The CN2 induction algorithm. Mach Learn 3:261–283
23. Cohen W (1995) Fast effective rule induction. In: Proceedings of the twelfth international conference on machine learning. Morgan Kaufmann Publishers Inc., San Francisco, pp 115–123
24. Cohen W, Singer Y (1999) A simple, fast, and effective rule learner. In: Proceedings of the sixteenth national conference on artificial intelligence and eleventh conference on innovative applications of artificial intelligence. AAAI Press, Palo Alto, pp 335–342
25. Covert I, Lee SI (2020) Improving kernelSHAP: practical Shapley value estimation via linear regression. Preprint. arXiv:201201536

26. Covert I, Lundberg S, Lee SI (2020) Understanding global feature contributions through additive importance measures. Preprint. arXiv:200400668
27. Crawford L, Flaxman S, Runcie D, West M (2019) Variable prioritization in nonlinear black box methods: a genetic association case study. Ann Appl Stat 13:958
28. Dembczyński K, Kotłowski W, Słowiński R (2008) Maximum likelihood rule ensembles. In: Proceedings of the 25th international conference on machine learning. ACM, New York, pp 224–231
29. Dembczyński K, Kotłowski W, Słowiński R (2010) ENDER: A statistical framework for boosting decision rules. Data Mining Knowl Discov 21:52–90
30. Devroye L, Wagner T (1979) Distribution-free inequalities for the deleted and holdout error estimates. IEEE Trans Inf Theory 25:202–207
31. Doshi-Velez F, Kim B (2017) Towards a rigorous science of interpretable machine learning. Preprint. arXiv:170208608
32. Dua D, Graff C (2017) UCI machine learning repository. http://archive.ics.uci.edu/ml
33. Efron B, Hastie T, Johnstone I, Tibshirani R (2004) Least angle regression. Ann Stat 32:407–499
34. Erhan D, Bengio Y, Courville A, Vincent P (2009) Visualizing higher-layer features of a deep network. University of Montreal 1341:1
35. Esposito F, Malerba D, Semeraro G, Kay J (1997) A comparative analysis of methods for pruning decision trees. IEEE Trans Patt Anal Mach Intell 19:476–491
36. Fokkema M (2017) PRE: An R package for fitting prediction rule ensembles. Preprint. arXiv:170707149
37. Freitas A (2014) Comprehensible classification models: A position paper. ACM SIGKDD Explorations Newsletter 15:1–10
38. Freund Y, Schapire R (1996) Experiments with a new boosting algorithm. In: Thirteenth international conference on ML, Citeseer, vol 96, pp 148–156
39. Friedman J (2001) Greedy function approximation: a gradient boosting machine. Ann Stat 29(5):1189-1232
40. Friedman J, Hastie T, Tibshirani R (2001) The elements of statistical learning, vol 1. Springer series in statistics. Springer, New York
41. Friedman J, Popescu B, et al. (2003) Importance sampled learning ensembles. J Mach Learn Res (2003) 4:94305
42. Friedman J, Popescu B, et al. (2008) Predictive learning via rule ensembles. Ann Appl Stat 2:916–954
43. Fürnkranz J (1999) Separate-and-conquer rule learning. Artif Intell Rev 13:3–54
44. Fürnkranz J, Widmer G (1994) Incremental reduced error pruning. In: Proceedings of the 11th international conference on machine learning. Morgan Kaufmann Publishers Inc., San Francisco, pp 70–77
45. Genuer R, Poggi JM, Tuleau-Malot C (2010) Variable selection using random forests. Patt Recogn Lett 31:2225–2236
46. Ghanem R, Higdon D, Owhadi H (2017) Handbook of uncertainty quantification. Springer, New York
47. Gregorutti B, Michel B, Saint-Pierre P (2017) Correlation and variable importance in random forests. Stat Comput 27:659–678
48. Guidotti R, Monreale A, Ruggieri S, Turini F, Giannotti F, Pedreschi D (2018) A survey of methods for explaining black box models. ACM Comput Surv 51:1–42
49. Guyon I, Weston J, Barnhill S, Vapnik V (2002) Gene selection for cancer classification using support vector machines. Mach learn 46:389–422
50. He Z, Yu W (2010) Stable feature selection for biomarker discovery. Comput Biol Chem 34:215–225
51. Iooss B, Lemaître P (2015) A review on global sensitivity analysis methods. Springer, Boston, pp 101–122

52. Iooss B, Prieur C (2017) Shapley effects for sensitivity analysis with correlated inputs: comparisons with Sobol'indices, numerical estimation and applications. Preprint. arXiv:170701334
53. Ish-Horowicz J, Udwin D, Flaxman S, Filippi S, Crawford L (2019) Interpreting deep neural networks through variable importance. Preprint. arXiv:190109839
54. Ishwaran H (2007) Variable importance in binary regression trees and forests. Electron J Stat 1:519–537
55. Ishwaran H, Kogalur U, Blackstone E, Lauer M (2008) Random survival forests. Ann Appl Stat 2:841–860
56. Kim B, Wattenberg M, Gilmer J, Cai C, Wexler J, Viegas F (2018) Interpretability beyond feature attribution: Quantitative testing with concept activation vectors (TCAV). In: International conference on machine learning, PMLR, pp 2668–2677
57. Kumar IE, Venkatasubramanian S, Scheidegger C, Friedler S (2020) Problems with shapley-value-based explanations as feature importance measures. In: III HD, Singh A (eds) Proceedings of the 37th international conference on machine learning, PMLR. Proceedings of machine learning research, vol 119, pp 5491–5500
58. Kumbier K, Basu S, Brown J, Celniker S, Yu B (2018) Refining interaction search through signed iterative random forests. arXiv:181007287
59. Letham B (2015) Statistical learning for decision making: interpretability, uncertainty, and inference. PhD thesis, Massachusetts Institute of Technology
60. Letham B, Rudin C, McCormick T, Madigan D (2015) Interpretable classifiers using rules and Bayesian analysis: Building a better stroke prediction model. Ann Appl Stat 9:1350–1371
61. Lipton Z (2016) The mythos of model interpretability. Preprint. arXiv:160603490
62. Liu S, Patel R, Daga P, Liu H, Fu G, Doerksen R, Chen Y, Wilkins D (2012) Combined rule extraction and feature elimination in supervised classification. IEEE Trans. Nanobiosci. 11:228–236
63. Louppe G (2014) Understanding random forests: From theory to practice. Preprint. arXiv:14077502
64. Lundberg S, Lee SI (2017) A unified approach to interpreting model predictions. In: Advances in neural information processing systems, New York, pp 4765–4774
65. Lundberg S, Erion G, Lee SI (2018) Consistent individualized feature attribution for tree ensembles. Preprint. arXiv:180203888
66. Malioutov D, Varshney K (2013) Exact rule learning via boolean compressed sensing. In: The 30th international conference on machine learning. Proceedings of machine learning research, pp 765–773
67. Meinshausen N (2010) Node harvest. Ann Appl Stat 4:2049–2072
68. Meinshausen N (2015) Package 'nodeharvest'
69. Mentch L, Hooker G (2016) Quantifying uncertainty in random forests via confidence intervals and hypothesis tests. J Mach Learn Res 17:841–881
70. Michalski R (1969) On the quasi-minimal solution of the general covering problem. In: Proceedings of the fifth international symposium on information processing. ACM, New York, pp 125–128
71. Murdoch W, Singh C, Kumbier K, Abbasi-Asl R, Yu B (2019) Interpretable machine learning: definitions, methods, and applications. Preprint. arXiv:190104592
72. Nalenz M, Villani M, et al. (2018) Tree ensembles with rule structured horseshoe regularization. Ann Appl Stat 12:2379–2408
73. Owen A (2014) Sobol'indices and Shapley value. SIAM/ASA J Uncertain Quant 2:245–251
74. Quinlan J (1986) Induction of decision trees. Mach Learn 1:81–106
75. Quinlan J (1987) Simplifying decision trees. Int J Man-Mach Stud 27:221–234
76. Quinlan J (1992) C4.5: Programs for machine learning. Morgan Kaufmann, San Mateo
77. Ribeiro M, Singh S, Guestrin C (2016) Why should I trust you? Explaining the predictions of any classifier. In: Proceedings of the 22nd ACM SIGKDD international conference on knowledge discovery and data mining. ACM, New York, pp 1135–1144
78. Rivest R (1987) Learning decision lists. Mach Learn 2:229–246

79. Rogers W, Wagner T (1978) A finite sample distribution-free performance bound for local discrimination rules. Ann Stat 6:506–514
80. Rüping S (2006) Learning interpretable models. PhD thesis, Universität Dortmund
81. Saltelli A (2002) Making best use of model evaluations to compute sensitivity indices. Comput. Phys Commun 145:280–297
82. Scornet E, Biau G, Vert JP (2015) Consistency of random forests. Ann Stat 43:1716–1741
83. Shah R, Meinshausen N (2014) Random intersection trees. J Mach Learn Res 15:629–654
84. Shapley L (1953) A value for n-person games. Contrib Theory Games 2:307–317
85. Shrikumar A, Greenside P, Kundaje A (2017) Learning important features through propagating activation differences. In: Proceedings of the 34th International Conference on Machine Learning. Proceedings of Machine Learning Research, pp 3145–3153
86. Simonyan K, Vedaldi A, Zisserman A (2013) Deep inside convolutional networks: visualising image classification models and saliency maps. Preprint. arXiv:13126034
87. Sobol I (1993) Sensitivity estimates for nonlinear mathematical models. Math Modell Comput Exp 1:407–414
88. Song E, Nelson B, Staum J (2016) Shapley effects for global sensitivity analysis: theory and computation. SIAM/ASA J Uncertain Quant 4:1060–1083
89. Song L, Smola A, Gretton A, Borgwardt K, Bedo J (2007) Supervised feature selection via dependence estimation. In: Proceedings of the 24th international conference on machine learning. Morgan Kaufmann Publishers, San Francisco, pp 823–830
90. Strobl C, Boulesteix AL, Zeileis A, Hothorn T (2007) Bias in random forest variable importance measures: illustrations, sources and a solution. BMC Bioinformatics 8:25
91. Su G, Wei D, Varshney K, Malioutov D (2015) Interpretable two-level boolean rule learning for classification. Preprint. arXiv:151107361
92. Sundararajan M, Najmi A (2020) The many Shapley values for model explanation. In: Thirty-seventh international conference on machine learning. Proceedings of machine learning research, pp 9269–9278
93. Tibshirani R (1996) Regression shrinkage and selection via the lasso. J R Stat Soc Ser B (Methodological), pp 267–288
94. Vapnik V (1998) Statistical learning theory. 1998, vol 3. Wiley, New York
95. Vaswani A, Shazeer N, Parmar N, Uszkoreit J, Jones L, Gomez A, Kaiser L, Polosukhin I (2017) Attention is all you need. Preprint. arXiv:170603762
96. Wager S, Athey S (2018) Estimation and inference of heterogeneous treatment effects using random forests. J Am Stat Assoc 113:1228–1242
97. Weiss S, Indurkhya N (2000) Lightweight rule induction. In: Proceedings of the seventeenth international conference on machine learning. Morgan Kaufmann Publishers Inc., San Francisco, pp 1135–1142
98. Williamson B, Feng J (2020) Efficient nonparametric statistical inference on population feature importance using Shapley values. In: Thirty-seventh international conference on machine learning. Proceedings of machine learning research, pp 10282–10291
99. Wright M, Ziegler A (2017) ranger: A fast implementation of random forests for high dimensional data in C++ and R. J Stat Softw 77:1–17
100. Yang H, Rudin C, Seltzer M (2017) Scalable bayesian rule lists. In: Proceedings of the 34th international conference on machine learning, PMLR, pp 3921–3930
101. Yu B (2013) Stability. Bernoulli 19:1484–1500
102. Yu B, Kumbier K (2019) Three principles of data science: predictability, computability, and stability (PCS). Preprint. arXiv:190108152
103. Zucknick M, Richardson S, Stronach E (2008) Comparing the characteristics of gene expression profiles derived by univariate and multivariate classification methods. Stat Appl Genet Mol Biol 7:1–34

Chapter 4
Interpretability in Generalized Additive Models

S. N. Wood, Y. Goude, and M. Fasiolo

Abstract Modelling the effect of a covariate vector on the distribution of a response variable, requires some structural assumptions, if the curse of dimensionality is to be avoided. Generalized additive models (GAMs) assume that the effects of the covariates are additive, with no, or only low order, interactions between effects. The additive assumption ensures scalability in the number of covariates and facilitates computational efficiency during model fitting. It also enhances model interpretability, which is critically important during model building and checking, as well as for communicating modelling results. This chapter formally introduces standard GAMs, as well as more flexibile GAMs for location shape and scale (GAMLSS). Is also shows how to interactively build and improve GAM and GAMLSS models via the mgcv and mgcViz R packages, which exploit their modular and interpretable structure. The final part of the chapter shows how to exploit the additive structure of GAMs to build powerful predictive models, by using random forests and online aggregation methods.

4.1 GAMs: A Basic Framework for Flexible Interpretable Regression

Suppose that we observe response variable y_i along side predictor variable vector x_i, where $i = 1, \ldots, n$, and that the aim is to learn how to predict future y values

S. N. Wood
University of Edinburgh, School of Mathematics, Edinburgh, UK
e-mail: simon.wood@ed.ac.uk

Y. Goude
EDF R&D, London, UK
e-mail: yannig.goude@edf.fr

M. Fasiolo (✉)
University of Bristol, School of Mathematics, Bristol, UK
e-mail: matteo.fasiolo@bristol.ac.uk

© The Author(s), under exclusive license to Springer Nature Switzerland AG 2022   85
A. Lepore et al. (eds.), *Interpretability for Industry 4.0 : Statistical and Machine Learning Approaches*, https://doi.org/10.1007/978-3-031-12402-0_4

from future observed **x** vectors. Note that in cases where i indexes time, $\mathbf{x}_i$ might include past values of y_i. In addition to predicting we might also be interested in understanding *how* y is related to **x**—to be able to *interpret* how the prediction process works. This can be important for sanity checking predictions, and often for guiding us as to what other predictors we might usefully incorporate in **x** (if yesterday's temperature was very important, would the day before's temperature also be worth considering, for example?).

Generically we want to learn the function f in the model

$$y_i = f(\mathbf{x}_i) + \text{'noise'}$$

or more generally the probability density function, π, in

$$y_i \sim \pi(y_i|\mathbf{x}_i).$$

Unless p, the dimension of **x**, is very low, then this problem suffers from the *curse of dimensionality*. For example, n^p uniformly spaced **x** points are required to cover the hypercube $[0, 1]^p$ as densely as n uniformly spaced x points could cover $[0, 1]$, and it is the density of points that determines the level of detail about f that we can hope to resolve. At the same time, even if we stick with a simple polynomial model type structure for f, we would need to use a polynomial of order at least p if we want to allow interactions involving all model terms (e.g., $x_1 x_2$ for $p = 2$, $x_1 x_2 x_3$ for $p = 3$, etc.). Such a model would have at least $(2p)!/(p!p!)$ parameters to estimate. To make progress we need to impose a more restrictive structure on the problem. Additive smooth models do this by allowing a flexible smooth dependence of y on each predictor, x_j, but assume additivity of effects, with no, or only low order, interactions between effects [16, 32]. For example if **A** is a model matrix, $\mathbf{A}_i$ is its i-th row and f_j are unknown smooth functions then a *generalized additive model* has the structure

$$g(\mu_i) = \mathbf{A}_i \gamma + \sum_j f_j(x_{ji}) \quad y_i \sim \text{EF}(\mu_i, \phi)$$

where g is a known smooth monotonic *link function* (identity, log, square root, etc.), EF denotes some exponential family distribution (Gaussian, Poisson, Gamma, bimomial etc.), γ is a parameter vector, ϕ a scale parameter and the f_j are subject to sum-to-zero identifiability constraints. The f_j need not be univariate: terms like $f_j(x_{ji}, x_{ki})$, $f_j(x_{ji}, x_{ki}, x_{li})$ etc. can be incorporated just as easily.

The model has a structured and interpretable form: functions of one, two and even three predictors are much easier to visualize than a general p dimensional function, for example. The curse of dimensionality is also much reduced by the decomposition into lower dimensional functions. The spacing of the predictor variables in p dimensional space is no longer the relevant consideration for how well we can estimate the model: the spacing of data in the low dimensional spaces relevant to each smooth term is now what matters. The price paid, of course, is the assumption that only main effects and some low order interactions matter. If that is not reasonable, then nothing may be gained. Notice however, that by specifying only that the f_j are smooth functions, the model also acquires a flexibility not shared by strictly parametric models.

4.1.1   Flexibility Can Be Important

Flexibility in the specification of the f_j can really matter in applications. For example Flaxman et al. [12] attempted to estimate the time course of the pathogen reproductive number R at the start of the Covid-19 epidemic in the UK by fitting a simple epidemic model to daily data on deaths from Covid. R was allowed to vary using a step function, with steps at each of 4 government policy change dates, the last one being lockdown. A Bayesian prior was used for the step heights which promoted fewer larger steps in preference to more smaller steps. Inference with this simple parametric model suggested that R was almost unchanged until full UK lockdown, when its value dropped to below 1. This corresponds to the infection rate continuing to surge until the eve of lockdown, thereafter collapsing.

However if the simple parametric step function is replaced with a flexible smooth function of the type considered here, allowing the data a greater role in determining the function shape, then the results change dramatically [33]. $R < 1$ before full lockdown, and new infections are in decline days earlier, as shown in Fig. 4.1. The latter result has since been confirmed by more direct methods (randomly sampled antibody positive subjects were asked when their symptoms started), and is also consistent with the dynamics that later occurred under less restrictive measures, similar to the situation pertaining before full lockdown.

The rather dramatic effect of relaxing strong parametric assumptions in this case is related to the highly non-linear nature of the model, whereby features of the data

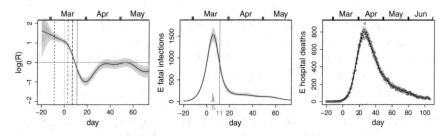

Fig. 4.1 Avoiding parameteric assumptions can make a big difference. Left: Log of inferred pathogen reproductive number, R, for England (2020), when it is assumed to be a smooth function of time: vertical lines show timing of social distancing interventions, with the red line being full lockdown. Middle: Corresponding inferred new (fatal) infections per day. Right: observed daily deaths (blue) and 100 forward simulations from inferred daily infection trajectory (grey). The original simple step function model for R in [12] implied surging infections and high R until the eve of lockdown. Subsequent data confirm the reconstruction shown here

that can not be captured with the parametric model (in particular inevitable variation in R post lockdown), result in compensating fitting artefacts that strongly effect timing. For the less non-linear model structures considered in this chapter such large effects are unlikely, but the example non-the-less serves as a warning of the potential consequences of parametric mis-specification.

4.1.2 Making the Model Computable

We need a representation of the model that can be computed with. The most fundamental components of this are a way of representing the f_j and a conventient way of measuring and controlling their smoothness. The representation part is achieved using spline type basis expansions

$$f_j(x) = \sum_{k=1}^{K} \beta_{jk} b_{jk}(x),$$

where the β_{jk} are unknown and the $b_{jk}(x)$ are simple functions chosen for nice approximation properties. K (which may vary with j, although we have not shown this notationally) is chosen to be large enough that we can be reasonably sure of avoiding the mis-specification bias that an overly restrictive model may cause. Ideally K is chosen to be somewhat larger than is likely to be necessary for representing f_j, but small enough that computation is not too onerous.

To avoid overfit with this generous choice for K we will need to control the smoothness of the f_j, by penalizing the model likelihood during estimation using wiggliness penalties, such as

$$\lambda_j \int f_j''(x)^2 dx = \lambda_j \boldsymbol{\beta}_j^\mathsf{T} \mathbf{S}_j \boldsymbol{\beta}_j.$$

The elements of $\mathbf{S}_j$ are known—being determined by the choice of basis functions and penalty. λ_j controls strength of penalization/smoothness of f_j—data driven selection of this parameter will be covered below. Many other choices of penalty are possible, all resulting in similar quadratic penalization of the likelihood.

A basis expansion used to fit a smooth curve, f, to data, under different degrees of penalization is shown in Fig. 4.2. A so called 'B-spline' basis is shown, but there are many other possibilities. Spline bases in general arise by mathematically working out the space of functions that minimise a particular penalty, subject to the constraint of interpolating some x, y data, or of approximating it to some specific level of accuracy. The basis shown is that for functions that minimize the penalty given above, while interpolating (or approximating) 30 x, y points, evenly spaced over the x range shown. The basis functions turn out to be made up of piecewise cubic polynomials continuous to second derivative, but having third derivative discontinuities at those 30 evenly spaced x values.

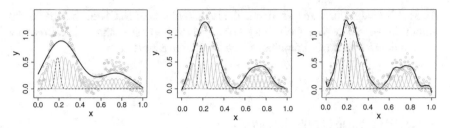

Fig. 4.2 Basis function expansions with different degrees of smoothness. Each panel shows the same set of basis functions multiplied by the corresponding coefficient in grey (i.e. $\beta_{jk} b_{jk}(x)$). The seventh is shown grey-black dashed to clearly illustrate the shape of a single basis function. The sum of the grey curves gives the black curve, which is a fit to the data shown as grey circles. The three panels show three fits, with successively higher values of the smoothing penalty, $\int f''(x)^2 dx$, allowed moving from left to right. The best fit is the middle panel, which avoids both over smoothing (left), and 'fitting the noise' (right)

S_j is typically not of full rank—for the given second derivative penalty example it is rank deficient by 2, the dimension of the space of functions linear in x (the functions that have zero penalty). This rank deficiency means that even with penalization, the f_j are typically not identifiable without constraint. Each f_j is only identifiable to within an intercept term. To deal with this an identifiability constraint is needed. A simple approach includes an overall intercept term in the model, but subjects each smooth to the constraint $\sum_{i=1}^{n} f_j(x_{ji}) = 0$ (or similar for functions of more than one predictor). This constraint is easily met by replacing each $b_{jk}(x)$ by

$$b_{jk}^*(x) = b_{jk}(x) - \sum_i b_{jk}(x_{ji})/n.$$

The basis now obviously spans a function space with 1 fewer degree of freedom than previously, but still with p basis functions—to remove this lack of independence we simply drop the basis function with the smallest value of $\sum_i b_{jk}^*(x_{ji})^2$ (so if the basis included the constant function—now reduced to the zero function— that's what gets dropped). Since the identifiability constraint is about removing constant functions from the basis for each smooth, the reparameterization does not change the meaning of the penalty. The constant function is in the null space of the penalty, so its removal does not alter the penalty, but of course under constraint the dimension of the penalty null space drops by one—for the penalty given above it would be one dimensional.

4.1.3 Estimation and Inference

Let us write all the GAM coefficients—γ and the basis coefficients for the f_j—in a single parameter vector β. Given the exponential family distributional assumption for y_i it is straightforward to compute the log likelihood of β. Given choices for the smoothing parameters λ_j controlling the weight to give to smoothness in estimation, we can then find the maximum penalized likelihood estimates

$$\hat{\beta} = \underset{\beta}{\text{argmin}} \; -l(\beta) + \beta^{\mathsf{T}} \mathbf{S}_\lambda \beta / 2$$

where $l(\beta)$ is the log likelihood and where

$$\mathbf{S}_\lambda = \begin{bmatrix} \lambda_1 \mathbf{S}_1 & 0 & \cdot \\ 0 & \lambda_2 \mathbf{S}_2 & \cdot \\ \cdot & \cdot & \cdot \end{bmatrix}.$$

This penalized likelihood can be justified in its own right, but it is also convenient to view it as arising from a Bayesian approach to smooth modelling, in which the penalty is induced by an improper Gaussian *smoothing prior*

$$\pi(\beta|\lambda) = N(0, \mathbf{S}_\lambda^-),$$

where $\mathbf{S}_\lambda^-$ is a pseudoinverse of $\mathbf{S}_\lambda$ (the precision matrix of the Gaussian prior). In that case $\hat{\beta}$ is obviously the posterior mode for β, but, in a regular large sample limit settings we also have the result that,

$$\beta|\mathbf{y}, \lambda \sim N(\hat{\beta}, (\hat{\mathcal{I}} + \mathbf{S}_\lambda)^{-1}) \tag{4.1}$$

where $\hat{\mathcal{I}}$ is the Hessian of the negative log likelihood, $-l$. This large sample Gaussian approximate posterior density will be denoted $\pi_g(\beta|\mathbf{y}, \lambda)$ below.

Approximation (4.1) can immediately be used to find credible intervals for β, or any quantity defined by β, such as an f_j. But the Bayesian view offers more, since it is also possible to infer appropriate values for the smoothing parameters, λ, in this framework. For example, we can find $\hat{\lambda}$ to maximize the Laplace Approximate Marginal Likelihood (LAML) $\pi_g(\mathbf{y}|\lambda) = \pi(\mathbf{y}|\hat{\beta})\pi(\hat{\beta}|\lambda)/\pi_g(\hat{\beta}|\mathbf{y}, \lambda)$ (so called because the expression is exactly the first order Laplace approximation to the marginal likelihood, $\int \pi(\mathbf{y}|\beta)\pi(\beta|\lambda)d\beta$).

The attentive reader will notice that these Bayesian computations have the same structure as those required for a generalized linear mixed model with Gaussian random effects—the smoothing parameters playing the part of inverse variance components. This duality between smooths and random effects means that many simple Gaussian random effect terms can be included in a GAM and treated just like the f_j, statistically and computationally.

Computation of $\hat{\lambda}$ is usually accomplished using a Newton or Quasi-Newton method, adapted to deal with the common case in which the optimal value for a λ_j is infinite, and using $\log \lambda_j$ as the working parameters, to keep λ_j positive. Computation of the LAML and its derivatives, required for optimization, involves recomputing the $\hat{\boldsymbol{\beta}}$ corresponding to each trial $\hat{\boldsymbol{\lambda}}$: an 'inner' Newton optimization is usually employed for this. The derivatives of $\hat{\boldsymbol{\beta}}$ w.r.t. to $\log \lambda_j$ are also required in order to compute derivatives of the marginal likelihood w.r.t. $\log \lambda_j$ for the outer optimization—implicit differentiation methods are usually used. More details can be found in [34] or [32], for example.

4.1.4   Checking, Effective Degrees of Freedom and Model Selection

Model checking with an exponential family GAM proceeds much as for a GLM. Similar residual checks are made to look for violations of the modelling assumptions, particularly violation of the assumed mean-variance relationship for y, and for any un-modelled auto-correlation in the residuals. Un-modelled auto-correlation in residuals ordered by a predictor can sometimes be an indicator that the basis dimension, K, used for the corresponding smooth term, was overly restrictive and a larger value is needed. The possibility that a K value is too small is generally something to bear in mind when model checking and is additional to the usual GLM checks.

One indicator that K *may* be too small is if the *effective degrees of freedom* of a smooth term is close to $K - 1$ (the -1 resulting from the identifiability constraint). Effective degrees of freedom attempt to capture the notion of how many unpenalized coefficients would result in a model term of equivalent flexibility to a given set of penalized coefficients, estimated under a particular level of penalization.

As a motivating example consider a smooth of one predictor, estimated with a squared second derivative penalty penalizing the likelihood, and having a basis dimension of 20. Under very high penalization (high λ_j) the smooth will be constrained to lie in the null space of the penalty—i.e. to be linear in the predictor. Despite its 19 coefficients the line has only one degree of freedom, corresponding to its slope (the constraint having removed the intercept). Under no penalization the smooth would obviously have 19 degrees of freedom, one for each free parameter. As we vary its smoothing parameter from infinity down to zero, the smooth becomes continuously less smooth and more complex, and it makes sense that for a level of smoothness somewhere between the straight line case and full unpenalized flexibility, the degrees of freedom should also be intermediate.

Consider the matrix $\mathbf{F} = (\hat{\mathcal{I}} + \mathbf{S}_\lambda)^{-1}\hat{\mathcal{I}}$ evaluated at $\hat{\boldsymbol{\beta}}$. This is the matrix approximately mapping the unpenalized $\hat{\boldsymbol{\beta}}$ to the penalized version, and so its diagonal elements, F_{ii}, can be viewed as 'shrinkage' factors for the coefficients. Summing up the elements F_{ii} corresponding to a single smooth term gives its *effective degrees of freedom* (EDF). In fact, since $\hat{\mathcal{I}}$ is not guaranteed to be positive definite at $\hat{\boldsymbol{\beta}}$, it is usually better to use the expected Hessian of the negative log likelihood in the computation of $\mathbf{F}$, to avoid occasional nonsensical EDF results. With this modification the EDF always decreases monotonically with increasing smoothing parameters and is bounded between $K - 1$ and the penalty null space dimension.

In addition to inference, such as credible interval computation, directly using the Bayesian result (4.1), some frequentist model selection tools are also useful. In particular it is possible to construct a generalized AIC

$$\text{AIC} = -2l(\hat{\boldsymbol{\beta}}) + 2\text{trace}(\mathbf{F}),$$

for model comparison, although for best performance this should incorporate a correction to the EDF (trace($\mathbf{F}$)) accounting for smoothing parameter uncertainty [34]. It is also possible to obtain approximate p-values for testing $H_0 : f_j = 0$ [30, 31]. Both approaches can provide useful guidance about which terms should be included in a GAM.

4.1.5 GAM Computation with mgcv in R

R package mgcv offers generalized additive modelling functions based on the preceding approach. In particular its gam function is used in a similar manner to the glm function for fitting generalized linear models in R. That is a response variable and linear predictor structure are specified using a *model formula*, while the distribution and link function are specified via a *family*, with the observations of the variables referred to in the model formula usually supplied in a *data frame*. To see this in action, consider a simple simulated Poisson example from the mgcv help files:

```
library(mgcv);set.seed(7)
dat <- gamSim(1,n=400,dist="poisson",scale=.2,verbose=FALSE)
b <- gam(y~s(x0)+s(x1)+s(x2),family=poisson,data=dat,method="REML")
par(mfrow=c(1,3),mar=c(4,4,1,1))
plot(b,scheme=1)
```

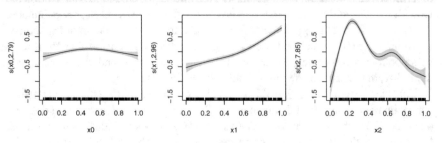

The model fitted here is

$$\log(\mu_i) = f_1(x_{0i}) + f_2(x_{1i}) + f_3(x_{2i}) \quad \text{where} \quad y_i \sim \text{Poi}(\mu_i)$$

and the plot shows the three estimated smooth functions with 95% credible intervals calculated using Equation (4.1). The numbers in the y axis labels such as s(x0,2.89) are the termwise effective degrees of freedom, while the ticks along the x axis show where observations occurred. Default basis dimensions of 10 have been used. s(x2,k=20) would have changed this to 20, for example.

Before taking credible intervals or model estimates seriously it is important to do some model checking of course. Some basic residual checks are provided as follows:

```
gam.check(b)
```

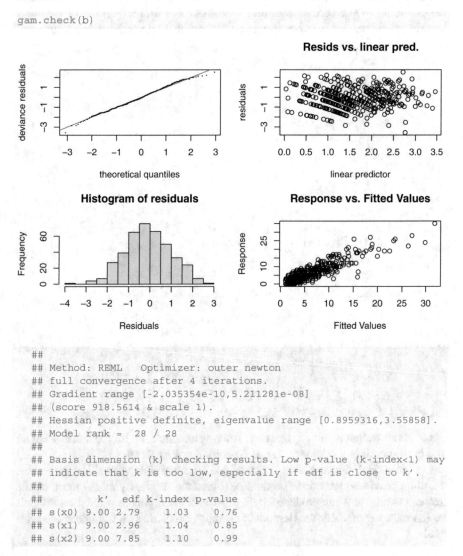

```
##
## Method: REML   Optimizer: outer newton
## full convergence after 4 iterations.
## Gradient range [-2.035354e-10,5.211281e-08]
## (score 918.5614 & scale 1).
## Hessian positive definite, eigenvalue range [0.8959316,3.55858].
## Model rank =  28 / 28
##
## Basis dimension (k) checking results. Low p-value (k-index<1) may
## indicate that k is too low, especially if edf is close to k'.
##
##         k'  edf k-index p-value
## s(x0) 9.00 2.79   1.03   0.76
## s(x1) 9.00 2.96   1.04   0.85
## s(x2) 9.00 7.85   1.10   0.99
```

The top right panel shows deviance residuals against the linear predictor—the residuals should show no pattern in mean or variance if the model structure is correct and the mean-variance model is appropriate. Top left is a qq-plot of the ordered residuals against theoretical quantiles—this should be close to a straight line if the distribution model is approximately correct. The lower two panels are self explanatory. The other output provides technical information on algorithm convergence, and an informal simulation test checking for residual correlation in the residuals when ordered against each predictor, which might indicate a problem with the basis dimension being too restrictive (although other issues may also cause an apparently significant result, so this is only a rough guide). The output here shows no problems. While serious checking should involve further plotting of residuals against predictors, let us move on, and consider model selection functions such as:

```
AIC(b)

## [1] 1802.089

summary(b)

##
## Family: poisson
## Link function: log
##
## Formula:
## y ~ s(x0) + s(x1) + s(x2)
##
## Parametric coefficients:
##             Estimate Std. Error z value Pr(>|z|)
## (Intercept)  1.58763    0.02481   63.99   <2e-16 ***
## ---
## Signif. codes:  0 '***' 0.001 '**' 0.01 '*' 0.05 '.' 0.1 ' ' 1
##
## Approximate significance of smooth terms:
##         edf Ref.df Chi.sq p-value
## s(x0) 2.787  3.467  14.19 0.00455 **
## s(x1) 2.956  3.670 404.51 < 2e-16 ***
## s(x2) 7.852  8.644 812.74 < 2e-16 ***
## ---
## Signif. codes:  0 '***' 0.001 '**' 0.01 '*' 0.05 '.' 0.1 ' ' 1
##
## R-sq.(adj) =  0.785   Deviance explained = 74.1%
## -REML = 918.56  Scale est. = 1        n = 400
```

Most of this output is self explanatory. The p-values are each for the test of whether the corresponding model term could be zero. Another standard task is prediction using the model. Given a data frame of predictor variable values, the predict function provides predictions on the linear predictor scale, by model term, on the response scale, or it can produce the "lpmatrix" mapping the model coefficients to the required predictions. Here is an example, predicting on the linear predictor scale:

```
pd <- data.frame(x0=c(.4,.1),x1=c(.3,.8),x2=c(.7,.1))
predict(b,newdata=pd,se=TRUE)

## $fit
##        1         2
## 1.299005  1.964851
##
## $se.fit
##         1          2
## 0.07988181  0.07898040
```

4.1.6   Smooths of Several Predictors

Now consider smooth model terms involving more than one predictor. Interpretability of model smooth terms involving multiple predictors requires some consideration to be given to how they are constructed. There are two main possibilities. One is to construct multidimensional analogues of smooths of one predictor, for which a single smoothing penalty treats smoothness in all predictor space directions isotropically. This is saying that if we travel along any two arbitrary lines in predictor space, we would expect the function to appear more or less equally smooth along both of them. The simplest example of such a penalty is the *thin plate spline* penalty for a smooth, $f(x_j, x_k)$, of two predictors,

$$\lambda \int \frac{\partial^2 f}{\partial x_j^2}^2 + 2\frac{\partial^2 f}{\partial x_j \partial x_k}^2 + \frac{\partial^2 f}{\partial x_k^2}^2 \, dx_j dx_k$$

It can be generalized to smooths of any number, d, of predictors and different orders, m, of derivative, although in general functions minimizing such a penalty while interpolating a given set of points only exist if $2m > d$ ($2m > d + 1$ if f should look visually smooth). A less intuitively interpretable penalty is required to restore complete freedom in the choice of m as d increases: the Duchon spline penalty, of which the thin plate spline penalty is a special case [9].

As in the case of the univariate penalty, given a penalty the basis for the space of functions minimizing the penalty while interpolating (or approximating to given accuracy) a set of points is known mathematically, and such a thin plate spline or Duchon spline basis can be used to represent f, exactly as in the univariate case. Again, given a basis with coefficients $\boldsymbol{\beta}$, the penalty has the form $\boldsymbol{\beta}^T S \boldsymbol{\beta}$, where S is fixed and known. In mgcv, terms like s(x1,x2,k=50) are used to add isotropic thin plate spline terms to the model. More predictors can also be used, and general Duchon penalty splines are also available.

Isotropic smoothing can also be generalized to applications such as smoothing on the surface of a sphere, as illustrated in the right panel of Fig. 4.3. However isotropic

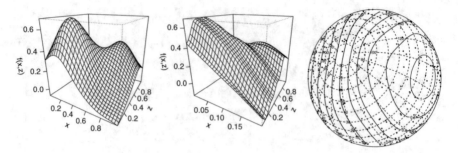

Fig. 4.3 Isotropy is a strong assumption, but isotropic smoothing is readily generalized. Left: A thin plate spline fitted to some data randomly and uniformly distributed over the unit x, z square. Middle: A thin plate spline fitted to the same data as in the left panel, but with the x co-ordinate divided by 5. The strong effect of the isotropic smoothness assumption is clearly visible: the smoother has the same level of variability per unit change in x as it has per unit change in z, but that results in a poor reconstruction here. Interaction/tensor product smooths do not suffer from this problem—they are in fact invariant to such linear rescalings. Right: an example of an isotropic smoother, where smoothness is defined over the surface of the sphere

smoothness is a strong assumption, as illustrated for a thin plate spline in the left two panels of the same figure.

The second approach is to abandon isotropy and instead use the notion of a statistical interaction to construct functions of several predictors (e.g. [29]). A statistical interaction occurs when the effect of one predictor is itself dependent on another predictor. Or more concretely, when the coefficients describing the effect of one predictor themselves depend on another predictor. Following this construction, we can construct an interaction spline of x_j and x_k by starting with a univariate spline basis expansion for the smooth of x_j,

$$f_j(x_j) = \sum_i \gamma_i g_i(x_j),$$

where the γ_i are coefficients and the g_i are basis functions. Now allow each coefficient γ_i to itself be a smooth function of x_k, using the univariate spline basis function expansion for smooths of x_k,

$$\gamma_i(x_k) = \sum_l \beta_{il} \delta_l(x_k),$$

where the β_{il} are coefficients and the δ_l are basis functions. Hence we arrive at

$$f(x_j, x_k) = \sum_i \sum_l \beta_{il} \delta_l(x_k) g_i(x_j).$$

So the basis functions for $f(x_j, x_k)$ are simply all the possible products of the basis functions for the univariate functions of x_j and x_k—for this reason the basis is often

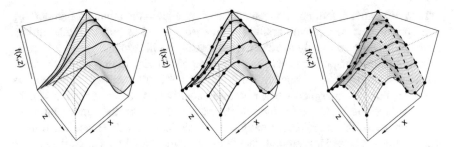

Fig. 4.4 Construction of an interaction (tensor product) smooth of z and x. Left: a smooth function of z is parameterized in terms of spline function values at 6 equally spaced z values, shown as black blobs. These parameters/coefficients are allowed to vary smoothly with x, as illustrated by the black curves, implying the surface represented by the grey curves. Middle: in practice the smooth variation of the coefficients with x is represented using spline functions of x, also parameterized in terms of function values at equally spaced x values, again shown as black blobs. Right: The construction is obviously symmetric. Smoothness of the surface $f(x, z)$ in the z direction can be measured by summing up the marginal penalties for smoothness in the z direction, for each of the dashed curves. Similarly smoothness in the x direction is measured by summing the marginal x direction penalty applied to the continuous black curves

known as a *tensor product* basis. The univariate bases used for this purpose are refered to as *marginal* bases, for obvious reasons. The construction is illustrated by Fig. 4.4, for marginal splines parameterized in terms of function values at equally spaced predictor values. Note that the construction is symmetric—the same basis arises if we start with the smooth of x_k and allow its coefficients to vary smoothly with x_j. It also generalizes immediately to more predictor variables. The coefficients of a smooth of two variables can be turned into smooth functions of a third variable, and so on.

What should we use as a smoothing penalty with such a tensor product basis? Interpretability is one important consideration. Another is that smoothing equally in all directions, as isotropic penalties do, is not really compatible with the notion of an interaction smooth. Suppose we are interested in how the effect of air temperature varies with elevation, for example. Why should we expect the same smoothness with respect to a unit change in temperature as we expect per unit change in elevation? And if we do make this choice the relative amount of smoothing with respect to elevation and temperature will be entirely dependent on the choice of measurement units, an essentially arbitrary decision.

A simple way to avoid such arbitrariness is to use a smoothing penalty that treats smoothness in different predictor variable directions separately, with at least one smoothing parameter per variable. For the example given above we might use

$$\lambda_j \int \left(\frac{\partial^2 f}{\partial x_j^2}\right)^2 dx_j dx_k + \lambda_k \int \left(\frac{\partial^2 f}{\partial x_k^2}\right)^2 dx_j dx_k. \tag{4.2}$$

With a different smoothing parameter for each predictor, the amount of smoothing appropriate w.r.t. to the two variables can be established as part of model fitting, rather than being arbitrarily fixed as part of model specification.

Computation of penalties like the one reported in Equation (4.2) is possible, but somewhat involved computationally, and a simpler alternative is equally interpretable. For any marginal basis a one-to-one linear reparameterization can be used to give the basis coefficients of the marginal smooth the interpretation of being values of the smooth at equally spaced predictor values. Hence when these coefficients are turned into smooth functions of another variable, e.g. x_k, they define smooth curves over the surface of f in the x_k direction. The whole collection of such curves can each be penalized by whatever interpretable penalty is associated with the marginal smooth for x_k. The right panel of Fig. 4.4 illustrates the idea. It then turns out that if $\gamma^\mathsf{T} \tilde{\mathbf{S}}_j \gamma$ was the smoothing penalty for the marginal smooth of x_j, then the x_j direction penalty for the tensor product has the form

$$\boldsymbol{\beta}^\mathsf{T} \mathbf{S}_j \otimes \mathbf{I} \boldsymbol{\beta},$$

where '$\otimes$' denotes the Kronecker product and $\mathbf{I}$ an identity matrix whose rank is the basis dimension for the x_k marginal smooth. The penalty in the x_k direction is $\boldsymbol{\beta}^\mathsf{T} \mathbf{I} \otimes \mathbf{S}_k \boldsymbol{\beta}$. Again, the construction generalized readily to more predictors.

In mgcv tensor product smooths using this construction are added to a model using model terms like te(x1,x2,x3). See ?te for more information.

4.1.7 Further Interpretable Structure

Viewing model smooth functions of more than one variable as statistical interaction terms leads naturally to structuring models in terms of 'main' smooth effects plus interactions, where the interaction now includes only smooth variation not captured by the main effects. For example we might be interested in smooth model terms of the form

$$f_j(x_j) + f_k(x_k) + f_{jk}(x_j, x_k),$$

where f_{jk} should now be constructed to exclude functions of the form $f_j(x_j) + f_k(x_k)$. Such a construction turns out to be very easy. Let's assume that the bases and penalties for f_j and f_k will be used as the marginal smooths for f_{jk} (although it's not strictly necessary to do so). If we remove the constant function from the basis for f_j, then we exclude the basis for the univariate effect f_k from the basis for f_{jk}, and eliminating the constant function from the f_k basis similarly removes the univariate f_j basis. So all we need do is to apply the sum-to-zero indentifiabilty constraints to the marginal bases (and penalties) *before* using them to construct f_{jk} and the resulting basis has the desired property, with penalties whose interpretation is unchanged (the resulting basis requires no further identifiability constraint, of course). In `mgcv` terms like `ti(x1,x2)` are used to construct such interactions without main effects (so something like `s(xj)+s(xk)+ti(xj,xk)` would implement the example above). Note that a term like `ti(x1,x2,x3)` excludes both main effects and lower order (i.e. second, here) interactions.

Of course we should not expect a model formulated this way to give identical results to one in which we do not specify main effects and interactions as separate smooths, and use a single tensor smooth constructed from unconstrained marginal bases. The space of functions defined by the model bases is identical in both cases, but the penalty is obviously not. There are 4 smoothing parameter in the main effects + interactions representation, rather than 2, for example.

Another common approach is to consider interactions of smooth and parametric effects. If the parametric term is the linear effect of a metric predictor variable then we get a so called *varying coefficient* term [17], such as $f(x_j)x_k$. Essentially the slope parameter associated with x_k varies smoothly with x_j (often time). When the smooth is a function of spatial location, e.g. $f(\text{lon}_i, \text{lat}_i)x_i$ then the model is sometimes refered to as 'geographic regression'. The obvious interaction of a smooth effect with a factor variable produces a separate smooth for each factor level—or alternatively one smooth for reference level of the factor, and then a 'difference' smooth for each remaining level.

Any linear functional of a smooth is also readily incorporated into the model. For example suppose that we are interested in the effect of temperature on electricity load over several preceding days. One possibility is to use terms like

$$\sum_{j=1}^{J} f_j(T_{i-j})$$

where each f_j is a separate smooth function and T_i is the mean temperature on day i. But is the effect of yesterday's temperature going to look entirely different to the effect of the day before yesterday's temperature? It seems more likely that the shape of the f_j would change smoothly with j. In that case we could instead use the model term

$$\sum_{j=1}^{J} f(T_{i-j}, j)$$

where f is a tensor product smooth of lagged temperature and the size of the lag. If `T` is an $n \times J$ matrix such that $T_{ij} = T_{i-j}$ and `L` is a matrix of the same dimension such that $L_{ij} = j$ then `te(T,L)` would produce just such a term in `mgcv`. The smooth is simply evaluated at each element of its matrix arguments, to produce a matrix of values, and the columns of this matrix are then summed over, to obtain the contribution to the model's linear predictor.

4.2   From GAM to GAMLSS: Interpretability for Model Building

As explained above, in standard GAM models the conditional distribution $\pi(y|x)$ is modelled via a distribution belonging to the exponential family with parameter vector $\theta = \{\mu, \phi\}$. While μ is allowed to vary with the covariates, ϕ does not vary with x. This setting can be generalized in two ways. First it is possible to consider distributions that do not belong to the exponential family, and that are parametrised by a general m-dimensional vector θ controlling (for example) the location, scale and skewness of the distribution. Second, it is possible model each element of θ via a separate additive model, that is

$$g^l\{\theta_i^l\} = \mathbf{A}_i^l \gamma^l + \sum_j f_j^l(x_{ji}), \quad \text{for } l = 1, \ldots, m. \tag{4.3}$$

Hence, we have l linear predictors, one for each parameter. The simplest example of a such a generalized additive models for location scale and shape (GAMLSS) model [27] is a Gaussian location-scale model where the mean, θ^1, and the variance, θ^2, are both allowed to vary with x. In mgcv package, such a model can be built and fitted to data via code such as

```
fit <- gam(formula = list(y ~ x1 + s(x2, k = 15, bs = "cr"),
                          ~ s(x2)),
           family = gaulss, data = SomeData)
```

Here the gaulss function implements the location-scale version of the Gaussian family. When this family is used, we have to provide gam with a list of formulae, the first of which is used to model the mean and the second the variance. Under the gaulss family, g_1 is by default the identity function while $g_2(\theta_2) = \log(\theta_2 + b)$, where $\theta_2 = 1/\sigma = \text{var}(y)^{-1/2}$ and b is a small positive constant useful to improve numerical stability. The mgcv package provides several other GAMLSS families covering, among others, the Tweedie (twlss), generalized extreme value (gevlss) and the zero inflated poisson (ziplss) distribution.

GAMLSS models are particularly useful when we want to model all the features of $\pi(y|x)$, rather than only the conditional mean $\mathbb{E}(y|x)$. This might be the case, for example, when forecasting electricity demand in the short term. In fact, the underlying economic loss $L(y, D)$ that characterises a trading or production planning decision D is not necessarily the quadratic loss, which is minimised by $\mathbb{E}(y|x)$. Hence, it is preferable to estimate the optimal decision D^*, by modelling the full distribution of $y|x$, using it to estimate $\mathbb{E}\{L(y, D)|x\}$ and minimising the latter w.r.t. D. Adequately modelling $\pi(y|x)$ might be important even for applications where $\mathbb{E}(y|x)$ is the only quantity of interest. This because the GAM fitting and inferential framework detailed in Sect. 4.1 relies on the model for $\pi(y|x)$ being at least approximately correct. Hence, for example, if the variance of y varies wildly with x, than the credible intervals obtained by fitting a Gaussian GAM with constant variance might have very poor frequentist coverage.

GAMLSS models provide extra modelling flexibility, but require more effort on the part of the modeller during model selection and checking. In fact, the user has to select which effects to include in each of the m linear predictors. Naïve automatic variable selection is often infeasible, due to the number of effects that could potentially be included and to the fact that GAMLSS models are computationally more expensive to fit than GAMs. If automated model building is key, an attractive option is the gamboostLSS R package, which implements the gradient boosting GAMLSS fitting methods proposed by Mayr et al. [21].

User-driven, interactive GAMLSS model development is complicated by inter-pretability issues. For instance, even a very experienced electricity demand modeller might not know which effects should be used to model the skewness of the demand distribution. To mitigate this issue, Fasiolo et al. [11] proposes a visualisation framework, implemented in the mgcViz package, which is meant to aid interactive GAMLSS model development and checking. In Sect. 4.2.1, we show how to use mgcViz for interactive model building, in the context of electricity demand modelling.

4.2.1   GAMLSS Modelling of UK Aggregate Electricity Demand

4.2.1.1   Data Overview and Pre-processing

This data set contains aggregate UK electricity demand data, at half-hourly resolution and covering the period from the 1st of January 2011 to the 30th of June 2016. The raw demand data was obtained from https://demandforecast.nationalgrid.com. Figure 4.5 shows some of the characteristics of the data. In particular, plot 4.5a shows that the long-term demand trend is negative. This is because the demand contained in the data set is net of embedded production from, e.g., solar panels and wind turbines. That is, net UK demand is decreasing because embedded generation is increasing faster than gross demand. The curves in plot 4.5b show the daily demand profiles for each day of the week. Unsurprisingly, demand is lower on weekends with a delayed morning peak, as people wake up later. The brightness of the hexagonal map in the background is proportional to the number of observations falling within each hexagon. Plot 4.5d shows how demand varies with temperatures, for each day of the week. The effect of temperature is similar on working days, while the heating effect ($t < 10°C$) seems stronger on Saturdays than on Sundays. Figure 4.5c shows that demand is higher during the winter than during the summer (time of year is 0 on Jan 1st and 1 on the 31st of Dec). Between years discrepancies in yearly demand profiles are substantial, especially in the winter.

We integrate the demand data with hourly temperatures, interpolated at the half-hourly resolution, from the NCEI. The temperature data was measured in the proximity of several large UK cities. We built a single temperature variable by

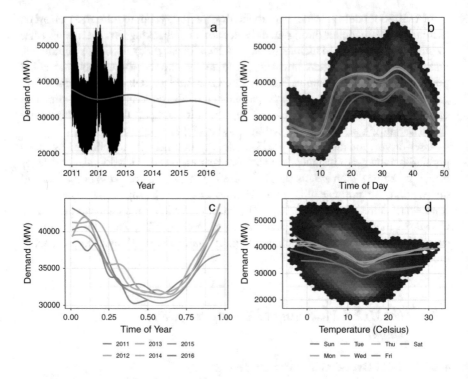

Fig. 4.5 The plots show: (**a**) demand vs time and the long-term trend; (**b**) the daily demand profiles for each day of the week; (**c**) seasonal demand dynamic for each year and (**d**) temperature effect for each day of the week. See the main text for more details

averaging these temperatures with weights that are proportional to the population of each city. Hence, the variable of interest are:

- `timeCount` is a progressive time counter;
- `toy` is the time of year from 0 (1st Jan) to 1 (31st Dec);
- `dow` is a factor variable indicating the day of the week;
- `holy` is a binary variable indicating holidays;
- `tod` is the time of day, ranging from 0 to 47, where 0 indicates the period from 00:00 to 00:30, 1 the period from 00:30 to 01:00 and so on;
- `temp` is the external temperature in degrees Celsius.
- `load48` the demand in the same half-hourly period of the previous day;
- `temp95` an exponential smooth of `temp`, that is `temp95[i]` = `a*temp[i]` + `(1-a)*temp95[i-1]` with `a = 0.05`.

The time period covered by the data set contains several exceptional days, on which demand cannot reasonably be modelled on the basis of historical data. In particular, we remove data corresponding to the 29th of April 2011 (Royal Wedding) and to the 4th–5th of June 2012 (Queen Elizabeth's Diamond Jubilee). We exclude also bank holidays and the days in the proximity of Easter, in particular: 22nd to 25 of April 2011, 6th to 9th of April 2012, 29th of March to 1st of April 2013, 18th to 21st of April 2014, 3rd to 6 of April 2015 and 25th to 28th of March 2016. The Christmas and New Year period is similarly difficult to forecast solely on the basis of historical data (that is, without manual or anyway ad hoc intervention), hence we remove data between 21st of December and the 4th of January of each year. Removal of these data leads to the vertical white stripes visible in Fig. 4.5a. We have not excluded data corresponding to the 2012 London Summer Olympic Games, but this is clearly an exceptional period as well.

4.2.1.2 Interactive GAMLSS Model Building

The data is contained in the `electBook` package, available at github.com/mfasiolo/electBook. We start by loading it:

```
library(electBook)
data(UKL)
```

The first model we consider is a simple Gaussian GAM model, with linear effects for the time trend, the lagged load and the day of the week. We include smooth effects for (smoothed) temperature, the time of day and the time of year. For the last two we use cyclical smooths (`bs = "cc"`). We use the `knot` argument to ensure that, for `toy`, the effect on the 1st of January matches the effect on the 31st of Dec while for `tod` the effect at 00:00 to 00:30 matches the same effect at the same time of the following day.

```
library(mgcv)
fit0 <- gam(load ~ timeCount + load48 + dow + s(temp) + s(temp95) +
              s(tod, bs = "cc") + s(toy, bs = "cc"),
          knots = list(tod = c(0, 48), toy = c(0, 1)), data = UKL)
```

To use the methods provided by `mgcViz` to visualise the fitted GAM, we need to load the package and to convert the output of `gam` to an object of class `gamViz`, via the `getViz` function:

```
library(mgcViz)
fit0 <- getViz(fit0, nsim = 100)
```

The arguments nsim is used to specify that we want to simulate 100 residuals vectors from the fitted GAM, which we use within the diagnostics offered by mgcViz. In particular, here we use check1D and l_gridCheck1D to check the residuals along the timeCount and load48 covariates:

```
pl0 <- check1D(fit0, list("timeCount", "load48", "toy")) +
        l_gridCheck1D(level = 0.95, showReps = FALSE)
print(pl0, pages = 1, layout_matrix = matrix(1:3, 1, 3))
```

The black points are the mean of the observed residuals from our model, binned depending on the value of the variable of the x axis. The red lines are 95% reference intervals, estimated by binning and averaging the simulated residuals, and computing the empirical quantiles of the binned means. The plots show that the residuals deviate significantly from what we would expect under the model. In particular, the plots suggest that the effects of timeCount and load48 are non-linear, while the fact that the mean residuals are negative for toy ≈ 0 and positive for toy ≈ 1 implies that using a cyclical smooth for toy is not appropriate.

We can check whether k is large enough via:

```
check(fit0)

...
##               k'   edf  k-index  p-value
## s(temp)     9.00  8.03    0.98     0.09  .
## s(temp95)   9.00  7.87    0.97     0.03  *
## s(tod)      8.00  8.00    0.96    <2e-16 ***
## s(toy)      8.00  7.72    0.94    <2e-16 ***
## ---
## Signif. codes:  0 '***' 0.001 '**' 0.01 '*' 0.05 '.' 0.1 ' ' 1
...
```

Similarly to gam.check, the code above produces some visual residual checks (not shown) and the table above. The table (see Sect. 4.1.5 for explanations on how to interpret it) suggests that we should consider using more basis functions for tod and toy. As a further check we can look at a QQ-plot:

```
qq(fit0, method = "tnormal", CI = "normal", level = 0.95)
```

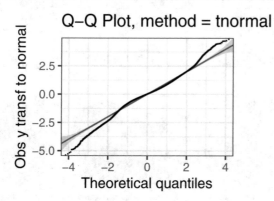

which shows that the residuals distribution has heavier tails than the Gaussian model used here. Hence, here we fit a more robust GAM based on a scaled Student's t distribution:

```
fit1 <- gam(load ~ dow + s(timeCount, k = 4) + s(load48) + s(temp) +
                s(temp95) + s(tod, k = 20, bs = "cc") + s(toy, k = 20),
            data = UKL, family = scat,
            knots = list(tod = c(0, 48)))
fit1 <- getViz(fit1, nsim = 100)
```

which also implement the other improvements we identified above, that is (a) smooth effects for timeCount and load48, (b) a larger basis dimension for tod and toy, (c) remove the cyclical constraint from s(toy).

With almost 10^5 observations, the UKL is large enough to consider including smooth interactions. The mgcViz package provides functions specifically designed to aid visual interaction detection. In particular, consider the following code:

```
pl <- check2D(fit1, x1 = list("load48", "temp", "temp95", "toy", "dow", "dow"),
                x2 = list("tod", "tod", "tod", "tod", "load48", "tod")) +
    l_gridCheck2D( )
print(pl + theme(legend.position="bottom"), pages = 1)
```

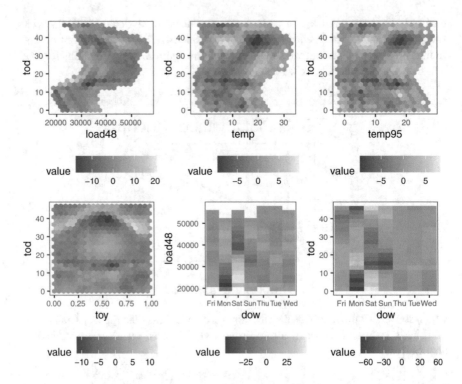

The heat maps are obtained by averaging the residuals falling within each hexagonal bin, and standardising the binned means via the binned simulated residuals. If the binned means are roughly normally distributed and the model is not missing any interaction, then we should expect most standardised means to fall in $(-2, 2)$, which is approximately a 95% Gaussian reference interval. But the colour scales show that the observed deviations are much larger than that. Further, the heat maps show some non-random patterns, which could be modelled via smooth interactions. The plots of `load48` and `tod` vs `dow` show that these effects depend on the day of the week. In particular, the plot on the bottom right suggests that the daily demand profile is different on Monday, Saturday and Sunday, relative to each other and to the rest of the week. To capture this in an improved model, we recode the `dow` variable as follows:

```
library(dplyr)
UKL$dowGroup <- recode_factor(UKL$dow, Tue = "Tue-Fri", Wed = "Tue-Fri",
                                        Thu = "Tue-Fri", Fri = "Tue-Fri")
levels(UKL$dowGroup)

   ## [1] "Tue-Fri" "Sun"        "Mon"       "Sat"
```

Checking again whether the bases dimensions are large enough:

```
check(fit1)

##
##                   k'     edf k-index p-value
## s(timeCount)    3.00    2.95    0.83  <2e-16 ***
## s(load48)       9.00    8.96    0.97   0.010 **
## s(temp)         9.00    7.48    1.00   0.605
## s(temp95)       9.00    7.45    0.99   0.310
## s(tod)         18.00   17.89    0.97   0.045 *
## s(toy)         19.00   17.57    0.93  <2e-16 ***
## ---
## Signif. codes:  0 '***' 0.001 '**' 0.01 '*' 0.05 '.' 0.1 ' ' 1
...
```

suggests that we should increase the value of k used for the effects of tod, toy, timeCount and load48. Increasing k for the first two effects is reasonable, but doing so for timeCount is risky, because the residuals of our model are auto-correlated. Residual auto-correlation is not modelled here, hence increasing k for the effect of timeCount would lead to a very wiggly effect, which extrapolates badly into the future. Hence, for this effect the value of k should not be chosen on the basis of the table above, but on other considerations. In particular, here the effect of timeCount is meant to capture long-term changes in load, due to technological and social changes that take place over the years. Hence, a small value of k is appropriate. We could consider increasing the value of k for load48 but, from an interpretability point of view, it is preferable to have an auto-regressive effect that is close to linear.

On the basis of the considerations above, we fit the following model:

```
fit2 <- bam(load ~ dow + s(timeCount, k = 4) +
                s(load48, k = 10, by = dowGroup, id = 1) +
                s(tod, k = 30, bs = "cc", by = dowGroup, id = 2) +
                s(temp, k = 20) + s(temp95, k = 20) +
                s(toy, k = 30) +
                ti(load48, tod, k = c(5, 5), bs = c("cr", "cc")) +
                ti(temp, tod, k = c(5, 5), bs = c("cr", "cc")) +
                ti(temp95, tod, k = c(5, 5), bs = c("cr", "cc")) +
                ti(toy, tod, bs = c("cr", "cc"), k = c(5, 5)),
            data = UKL,
            family = scat,
            knots = list(tod = c(0, 48)),
            discrete = TRUE)
fit2 <- getViz(fit2, nsim = 100)
```

The model is fitted via the bam function, which implements the big data GAM methods of [35]. In particular, setting discrete = TRUE leads to faster computation via covariate discretisation. In s(load48, k = 10, by = dowGroup, id = 1) the by argument allows us to create smooth effects of load48 that are different depending on the value of the dowGroup factor. By setting id = 1 we impose that wiggliness of the by-factor smooths should be penalised via a single smoothing parameter, rather than one for each factor level.

Let us look at all the by-dowGroup effects:

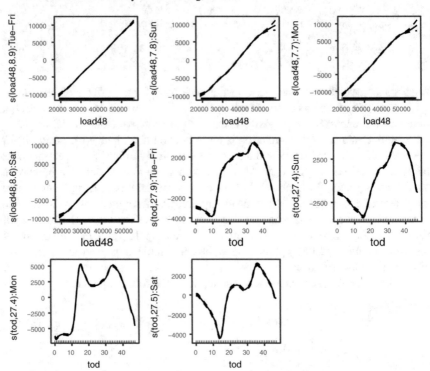

The effect of tod is clearly highly dependent on dowGroup. This is not surprising, for instance most people wake up later on Sunday than on working says, hence the morning demand ramp is gentler on Sundays. It is hard to visually assess whether the effect of load48 is different depending on dowGroup. The difference can be visualised via the following plot:

```
plotDiff(sm(fit2, 3), sm(fit2, 5)) + ylab("s(load48|Sun) - s(load48|Sat)")
```

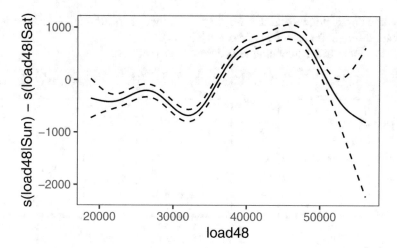

Under the `dowGroup` grouping, the weekdays between Tuesday and Friday are grouped together hence our model assumes that the effects of `load48` and `tod` do not vary strongly between those days. We can check whether this is the case via a further interaction check:

```
pl <- check2D(fit2, list("dow", "dow"), list("load48", "tod")) + l_gridCheck2D()
print(pl + theme(legend.position="bottom"), pages = 1)
```

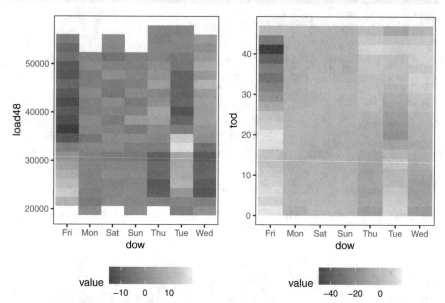

The left plot suggests that the effect of `load48` is significantly different depending on the day of the week. Strong residuals pattern are visible on Friday and Tuesday. The right plot show that we are grossly over-estimating the demand on Friday night. This suggest that grouping together the days between Tuesday and Friday might not

have been a great idea, and that we might be better of specifying by-factor effects
that are different for each day of the week (dow).

Let us looks at the temperature and time of year effects:

```
print(plot(fit2, select = 10:12), pages = 1, layout_matrix = matrix(1:3, 1))
```

The plot for toy shows a sharp trough in the summer (toy≈ 0.6) and a narrow
peak in late winter. In particular, the effect seems to be more wiggly in some
periods than in others. We could accommodate for this by using an adaptive smooth
(bs = "ad"), under which the smoothness penalty varies with the variable of
interest (toy here).

The following model:

```
fit3 <- bam(load ~ dow + s(timeCount, k = 4) +
                   s(load48, k = 10, by = dow, id = 1) +
                   s(tod, k = 30, bs = "cc", by = dow, id = 2) +
                   s(temp, k = 20) + s(temp95, k = 20) +
                   s(toy, bs = "ad", k = 30) +
                   ti(load48, tod, k = c(5, 5), bs = c("cr", "cc")) +
                   ti(temp, tod, k = c(5, 5), bs = c("cr", "cc")) +
                   ti(temp95, tod, k = c(5, 5), bs = c("cr", "cc")) +
                   ti(toy, tod, bs = c("cr", "cc"), k = c(5, 5)),
                   data = UKL,
                   family = scat,
                   knots = list(tod = c(0, 48)),
                   discrete = TRUE)
fit3 <- getViz(fit3, nsim = 100)
```

integrates the improvement identified above, namely an adaptive smooth for toy,
and by-dow smooths for load48 and tod. The model achieves the lowest AIC so
far:

```
AIC(fit0, fit1, fit2, fit3)

##                df       AIC
## fit0    41.62090 1584219
## fit1    73.20387 1569250
## fit2   261.01765 1432123
## fit3   370.42144 1418360
```

One could consider improving it further, for example by checking whether the number of basis function used to buing the tensor product interactions is sufficiently large. However, below we assume that the last GAM above models the conditional mean demand sufficiently well and we move now to the task of modelling the whole conditional demand distribution in a GAMLSS framework.

We start by looking at how the standard deviation of the residuals varies with some of the covariates:

```
pl0 <- check1D(fit3, list("dow", "load48", "temp", "temp95", "tod", "toy")) +
  l_gridCheck1D(gridFun = sd, level = 0.95, showReps = FALSE)
print(pl0, pages = 1)
```

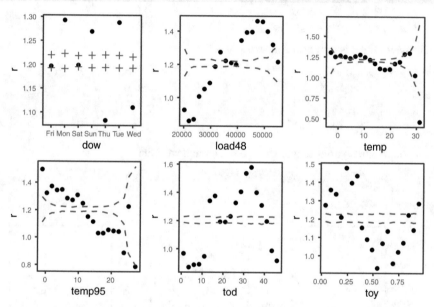

The plots are obtained by binning observed and simulated residuals as before, but now we are summarising the binned residuals via sd, rather than mean. The plots show that the residual variance varies wildly with most covariates. Importantly, the variance changes smoothly with the continuous covariates, which suggests that we could model these patterns via smooth effects.

To model the variance as well as the mean, we consider the following Gaussian location-scale model:

```
fit_lss <- gam(list(load ~ dow + s(timeCount, k = 4) +
                    s(load48, k = 10, by = dow, id = 1) +
                    s(tod, k = 30, bs = "cc", by = dow, id = 2) +
                    s(temp, k = 20) + s(temp95, k = 20) +
                    s(toy, bs = "ad", k = 30) +
                    ti(load48, tod, k = c(5, 5), bs = c("cr", "cc")) +
                    ti(temp, tod, k = c(5, 5), bs = c("cr", "cc")) +
                    ti(temp95, tod, k = c(5, 5), bs = c("cr", "cc")) +
                    ti(toy, tod, bs = c("cr", "cc"), k = c(5, 5)),
                  ~ dow + s(load48) + s(temp95) + s(tod, k = 20) + s(toy)),
               data = UKL,
               family = gaulss,
               knots = list(tod = c(0, 48)))
fit_lss <- getViz(fit_lss, nsim = 100)
```

The new model is preferable in terms of AIC:

```
AIC(fit3, fit_lss)

##                 df       AIC
## fit3     370.4214 1418360
## fit_lss  419.7592 1410087
```

and the residual variance patterns are now much weaker:

```
p10 <- check1D(fit_lss, list("dow", "load48", "temp", "temp95", "tod", "toy")) +
  l_gridCheck1D(gridFun = sd, level = 0.95, showReps = FALSE)
print(p10, pages = 1)
```

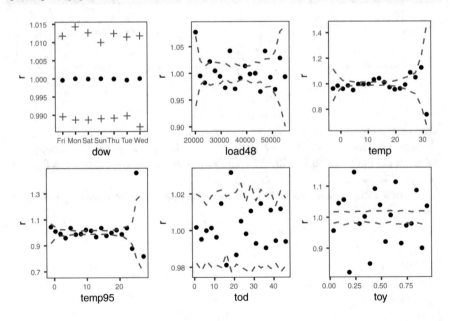

As we did for the mean model earlier, we could now verify whether the number of basis functions used to model the effects in the variance model is sufficiently large and we could look for bivariate interactions. While we leave this to the interested reader, here we verify whether further features of the conditional response distribution are correctly captured by the model. In particular, we look at how residual skewness varies with the covariates:

```r
library(e1071) # Provides the skewness() function
pl0 <- check1D(fit_lss, list("dow", "load48", "temp", "temp95", "tod", "toy")) +
    l_gridCheck1D(gridFun = skewness, level = 0.95, showReps = FALSE)
print(pl0, pages = 1)
```

Interestingly, the plots show that, on average, the residuals are skewed to the left. We see this from the fact that most observed binned skewness values fall below zero. Of course, skewness can not be model via a Gaussian distribution, which is symmetric. Further, the skewness varies with most covariates. For instance, the residual distribution is left skewed on Mondays and Fridays, and in the mornings (tod≈20). The skewness patterns are less smooth than the variance patterns observed above, hence we can not expect smooth effects to capture them as well as they did for the variance.

To model the demand skewness, we consider the four-parameter sinh-arcsinh distribution of [19]. The model allows us to control the location, scale, asymmetric and tail behaviour of the distribution via separate parameters. While we could have a full additive model for each, here we consider the following model:

```
fit_shash <- gam(list(load ~ dow + s(timeCount, k = 4) +
                    s(load48, k = 10, by = dow, id = 1) +
                    s(tod, k = 30, bs = "cc", by = dow, id = 2) +
                    s(temp, k = 20) + s(temp95, k = 20) +
                    s(toy, bs = "ad", k = 30) +
                    ti(load48, tod, k = c(5, 5), bs = c("cr", "cc")) +
                    ti(temp, tod, k = c(5, 5), bs = c("cr", "cc")) +
                    ti(temp95, tod, k = c(5, 5), bs = c("cr", "cc")) +
                    ti(toy, tod, bs = c("cr", "cc"), k = c(5, 5)),
                  ~ dow + s(load48) + s(temp95) + s(tod, k = 20) + s(toy),
                  ~ dow + s(load48) + s(temp95) + s(tod) + s(toy),
                  ~ 1),
             data = UKL,
             family = shash,
             knots = list(tod = c(0, 48)))
fit_shash <- getViz(fit_shash, nsim = 100)
```

which assumes that the fourth parameter, which controls the tail behaviour, does not depend on the covariates. One could consider checking whether the kurtosis of the residuals varies with the residuals, but we leave this to the interested reader. This final model is the best in terms of AIC:

```
AIC(fit_lss, fit_shash)

##                 df      AIC
## fit_lss    419.7592 1410087
## fit_shash  462.7579 1403485
```

It is interesting to visualise the fitted effects in the scale and skewness models:

```
print(plot(fit_shash, select = 23:30), pages = 1)
```

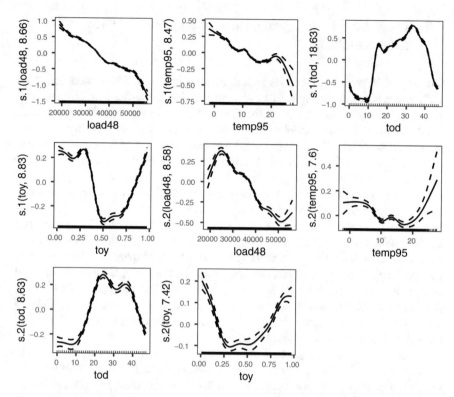

The effects on the variance are labelled s.1 and those on the skeweness s.2. Some effects are quite interpretable. For example, the plots for s.1(tod) and s.1(toy) show that the variance is minimal during the night and in the summer, which is reasonable. The plots on for s.2(tod) and s.2(toy) show that the demand is skewed to the right in the middle of the day and in the winter, which is also expected. It is more difficult to interpret the effects of lagged demand and of smoothed temperature of conditional variance and skewness.

In this section we considered an electricity demand forecasting application and we explained how to construct increasingly complex GAMLSS models in an interpretable manner, by exploiting the visualisation tools provided by mgcViz. However, electricity demand dynamics are complex and ever-changing, due to socio-economical factors and technological change, hence even a carefully crafted GAMLSS model might not be able to model electricity demand in all circumstances. The issue is considered in the next section, where we explain how additive and other types of models can be combined to produce an improved forecasting model, while retaining a degree of interpretability.

4.3   From GAMs to Aggregations of Experts, Are We Still Interpretable?

One important concern in time series forecasting and in particular in electricity load forecasting is due to changes in $\pi(y|x)$ with time, that is the data are not i.i.d.. Examples of such changes can be found recently due to the COVID 19 pandemic where a lot of human activites were considerably affected: mobility [24, 25], economic activities [22], pollution [20] and energy consumption [18]. It entailed new challenges in the electricity demand forecasting field. A recent forecasting competition [10] has been set up where the participants faced with the problem of forecasting electricity load consumption during the COVID lock-down in a big city (unknown location). Two of the three top teams [8, 36] applied online aggregation of experts approaches. Online robust aggregation of experts [6, 7] is a powerful model free approach aiming at learning in a streaming fashion characteristics of time series data. It consists in combining forecasts (called experts) according to their past performances. The methods are designed to be efficient even in an adversarial setting, that is where the data generating process can come from an adversary, making them very robust. In a changing context like the one induced by COVID 19 pandemic, online aggregation allows to adapt to changes in distribution and to track the performance of the best experts.

To produce good aggregated forecasts, it is important to aggregate "diverse" experts, coming for various methods and/or using different sources of information [13]. More generally, in the ensemble methods frameworks (bagging, random forest, boosting, etc.), it is well known that increasing the diversity of the base learners improves the ensemble performance and this can be done implicitly by choosing complementary models, learning models on different subsets/transformation of the data (bootstrapping, changing the temporal or spatial resolution, etc.) or optimizing different loss functions. It can also be done explicitly by optimizing some diversity criteria to force base learners to be diverse [2, 26].

We will focus here on experts generated with two methods that are quite complementary: GAMs and Random Forests (RFs). Whereas RFs are algorithmic based, GAMs are more model oriented, as explained above. GAMs performs well on electricity load data and have the nice property to be understandable by humans, the latter property being of crucial interest for the energy production and distribution planning. The price to pay for this interpretability is that designing a good GAM model requires some statistical background as well as domain specific knowledge. In contrast, RFs are powerful black box methods for modelling complex regression relationships [3] with very little prior knowledge of the problem. A RF regression aims at fitting a generic relationship $y_t = f(\mathbf{x}_t) +$ 'noise' by using an efficient and easy to tune data driven optimization algorithm. RFs are obtained by aggregating an ensemble of base learners generated by applying classification and regression trees [4] on different subsets of the data, obtained with bagging and random sampling of covariates. RFs capture well complex non-linear interactions, can be easily adapted in a time varying environment but their forecasts are by nature restricted to the convex envelop of the training data. However, GAMs allow to explicitly control the smoothness of the functional relationships and have good extrapolation properties. A nice compromise between the two approaches can be obtained by stacking them, as done in [14].

To stack a GAM and a RF we proceed in three steps:

1. Fit a GAM model to the training set and extract the estimated effects $\widehat{f}_j(x_j)$. We denote $\widehat{\mathbf{f}}(\mathbf{x})$ the set of these effects.
2. Estimate the forecasting residuals (either by cross-validation, block cross-validation or forecasting errors in an online forecasting setting) denoted $\widehat{\varepsilon}_t$.
3. Fit a RF model to predict the GAM residuals: $\widehat{\varepsilon}_t = g(\mathbf{x}_t, \widehat{\mathbf{f}}(\mathbf{x})) + u_t$. Hence, the set of features used by this RF includes the original features and the evaluated GAM effects. The final forecasts are obtained summing the GAM forecasts and the corrections provided by the RF.

To fit the RFs, we use the `ranger()` function from the `ranger` R package. The default parameters are used (in particular, 500 trees, $mtry = \sqrt{p}$, unlimited tree depth). For the GAMs we use `mgcv`. We also try a variant of RF presented in [15], using moving block bootstrap as an alternative to the classical bootstrap step of the RF algorithm. This is implemented in the `rangerts` package available at https://github.com/hyanworkspace/rangerts.

4.3.1 Online Forecasting with Online Aggregation of Experts

Here we present the experts that we will use in the aggregation and how to compute them on the UK dataset described in Sect. 4.2.1.1. We choose to focus on a subset of this dataset corresponding to `tod` = 40, i.e. 8 pm. We do that for two reasons: to reduce computational time but also because fitting the models and the aggregation to each half-hour of the day usually improve the forecasting performance. We split the data in two, the observations before the 1st of September 2015 are used to train the experts, while the following 281 days are used to the evaluate the online aggregation:

```
library(electBook)
data(UKL)
selH <- which(UKL$tod==40)
UKL_h <- UKL[selH, ]

n0 <- which(as.character(UKL_h$date)=="2015-09-01 20:00:00")
UKL_h0 <- UKL_h[1:n0, ]
UKL_h1 <- UKL_h[-c(1:n0), ]
```

The first expert, called g0, is inspired by the simplest GAM model from Sect. 4.2.1.2:

```
g0 <- gam(load ~ s(timeCount, k=4) + load48 + dow + s(temp) + s(temp95) +
                 s(tod, bs = "cc") + s(toy, bs = "cc"),
          knots = list(tod = c(0, 48), toy = c(0, 1)), data = UKL)
g0.forecast <- predict(g0, newdata=UKL_h1)
```

The second one, called g1, is obtained using a similar equation, but fitted on the single instant 40:

```
g1 <- gam(load ~ s(timeCount, k=4) + load48 + dow + s(temp) + s(temp95) +
                 s(toy, bs = "cc"),
          knots = list( toy = c(0, 1)), data = UKL_h0)
g1.forecast <- predict(g1, newdata=UKL_h1)
```

The third and fourth experts are RFs fitted to data from 8 pm only, with the same covariates the GAMs. rf0 is a basic RF and rf1 is a moving block bootstrap variant with a block size of 3 weeks:

```
library(ranger)
library(rangerts)
set.seed(350)
rf0 <- ranger(load ~ timeCount + load48 + dow + temp + temp95 + toy,
              data = UKL_h0)
rf0.forecast <- predict(rf0, data=UKL_h1)$prediction

rf1 <- rangerts(load ~ timeCount + load48 + dow + temp + temp95 + toy,
                data = UKL_h0, bootstrap.ts= "moving", block.size=7*3)
rf1.forecast <- predict(rf1, data=UKL_h1)$prediction
```

The fifth expert is the RF stacking model, using the GAM effects of g1 as supplementary covariates. We call it rfgam:

```
terms0 <- predict(g1, newdata=UKL_h0, type='terms')
terms1 <- predict(g1, newdata=UKL_h1, type='terms')
colnames(terms0) <- paste0("gterms_", c(1:ncol(terms0)))
colnames(terms1) <- paste0("gterms_", c(1:ncol(terms1)))

Data_rf0 <- data.frame(UKL_h0, terms0)
Data_rf0$res <- UKL_h0$load-g1$fitted.values

Data_rf1 <- data.frame(UKL_h1, terms1)
Data_rf1$res <- UKL_h1$load-g1.forecast

cov <- "timeCount + load48 + dow + temp + temp95 + tod + toy +"
gterm <-paste0("gterms_", c(1:ncol(terms0)))
gterm <- paste0(gterm, collapse='+')
cov <- paste0(cov, gterm, collapse = '+')
formule_rf <- paste0("res", "~", cov)
rfgam<- ranger(formule_rf, data = Data_rf0, importance = 'permutation')
rfgam.forecast <- predict(rfgam, data=Data_rf1)$prediction + g1.forecast
```

Table 4.1 Performance of the experts and of the online aggregation

	g0	g1	rf0	rf1	rfgam	best convex	BOA
MAPE	2.09	1.14	1.58	1.59	1.06	1.05	1.03
RMSE	1012.70	590.57	818.28	823.62	557.94	555.06	546.28

We can then compute an online aggregation of experts with the subgradient version of Bernstein Online Aggregation (BOA) algorithm [28] as well as the oracle corresponding to the a posteriori best convex aggregation.

```
library(opera)
experts <-  cbind(g0.forecast, g1.forecast, rf0.forecast, rf1.forecast,
                  rfgam.forecast)
colnames(experts) <- c("g0", "g1", "rf0", "rf1", "rfgam")
agg <- mixture(Y=UKL_h1$load, experts=experts, model='BOA',
               loss.gradient=T)
```

Table 4.1 presents the forecasting performances of the experts and of BOA on the test set, showing the improvement induced by fitting the GAM to data from 8 pm and the further performance improvement obtained via stacking RF and GAM.

As expected, the aggregation performs well, if we consider the performance of the best expert and the best convex aggregation. Figure 4.6 is generated by (we recall that the different experts are: g0 (simplest GAM), g1 (GAM by instant), rf0 (basic random forest), rf1 (block bootstrapped random forest), rfgam (stacked GAM and random forest)):

```
plot(agg, type='plot_weight', dynamic = F)
```

and it represents the evolution of the weights on the test set, and clearly shows the substantial weight given to the stacking forest as well as the dynamic evolution of the weights as functions of time.

4.3.2 Visualizing the Black Boxes

The problem of explaining black box machine learning models has received much attention in past few years [23]. Here we focus on accumulated local effects (ALEs) [1], which are popular methods to visualize black boxes by framing them as additive models. ALE methods is implemented in the ALEplot R package which can be used to visualise online aggregations, if a corresponding predict function is provided.

Figure 4.7 represents the ALEs associated with the covariates timeCount, load48, temp and toy (note that the timeCount effect overlaps for g0 and g1, as well as for rf0 and rf1). Notice that ALEs allow us to interpret clearly the effects of covariates on the different models as well as on the aggregation. Regarding the timeCount effect, this is constant for rf0 and rf1, which is due to the fact

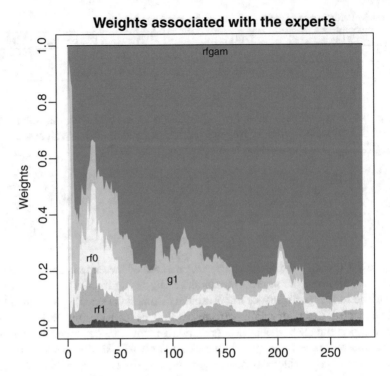

Fig. 4.6 Weights of the BOA algorithm on the test set

that the RF extrapolation is in the convex hull of the estimation data. This is not the case for the GAM models where we clearly see the linear negative extrapolation trend of g0 and g1. Given that the effects of g1 are added to rfgam, this RF can extrapolate as a GAM model. The ALE of timeCount on the aggregation is not linear, due the online adjustment of the weights. We also see that most of the time the aggregation trend falls in the convex hull of the other experts' trends.

Regarding the effect of load48, we see two groups of effects: the GAM ones (including rfgam), which are linear, and the RF ones which are close to piecewise linear functions. The ALE on the aggregation does not fall in the convex hull of the experts' ALEs, but it is closer to RF ones than to GAM ones. For the temp effect, all the models obtain similar effects for low temperatures (even if the GAM ones are again smoother than the RF ones), which is due to the impact of electrical heating. We observe some important differences for high temperatures between the models. This temperature effect has to be interpreted jointly with the toy effect, due to the yearly seasonality of the temperature. We observe very different yearly patterns, in particular during the summer holidays period when g1 and rfgam clearly see a decrease of the load whereas the RFs and g0 effects are quite flat. Further, we see a gap in the RF effects between the beginning of the year and the end of the year, whereas the cyclic constraints included in the GAMs do not allow such a

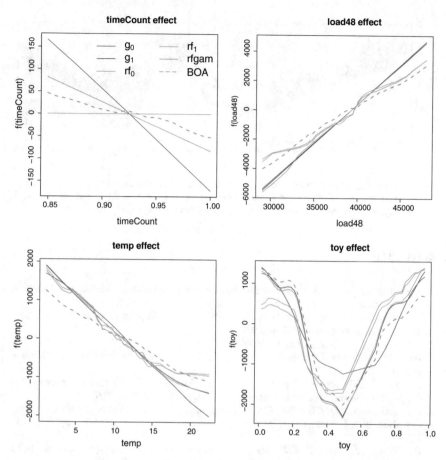

Fig. 4.7 Accumulated local effects for 4 covariates, for all our ML models and the online aggregation over the test set

discontinuity. Again, BOA produces effects similar but adapted to the test data for both temperature and yearly effects.

The ALE representation can be used also in an online setting, to visualize the time evolution of the different effects embedded into the aggregation. We plot these time varying ALEs for `temp` and `toy` in Fig. 4.8. We clearly see how the aggregation effects are learned sequentially. We also see that the domain of the effects (the x-axis range) evolves with time.

This experiment shows that we can adapt ALE plots to the context of online aggregation of experts and interpret it as an additive model, making it more explainable and thus increasing trust in the models. Another alternative could be to introduce additive constraints in the aggregation algorithm. This has been proposed by Capezza et al. [5], but not in an online setting.

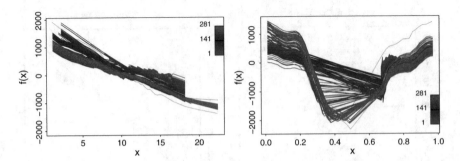

Fig. 4.8 Time varying accumulated local effects for temperature and time of year effects extracted from BOA aggregation (from the first instant of the test set in blue to final day in red). Dotted black lines correspond to the convex hull of the ALE effects of the individual experts

References

1. Apley DW, Zhu J (2020) Visualizing the effects of predictor variables in black box supervised learning models. J R Stat Soc Ser B (Stat Methodol) 82(4):1059–1086
2. Bourel M, Cugliari J, Goude Y, Poggi JM (2020) Boosting diversity in regression ensembles. https://hal.archives-ouvertes.fr/hal-03041309/
3. Breiman L (2001) Random forests. Mach Learn 45(1):5–32
4. Breiman L, Friedman J, Olshen R, Stone C (1984) Classification and regression trees. Chapman & Hall/CRC
5. Capezza C, Palumbo B, Goude Y, Wood SN, Fasiolo M (2021) Additive stacking for disaggregate electricity demand forecasting. Ann Appl Stat 15(2):727–746
6. Cesa-Bianchi N, Lugosi G (2006) Prediction, learning, and games. Cambridge University Press
7. Cesa-Bianchi N, Orabona F (2021) Online learning algorithms. Annu Rev Stat Appl 8(1):165–190
8. De Vilmarest J, Goude Y (2021) State-space models win the IEEE dataport competition on post-covid day-ahead electricity load forecasting. Tech. rep., arXiv:2110.00334
9. Duchon J (1977) Splines minimizing rotation-invariant semi-norms in Solobev spaces. In: Schemp W, Zeller K (eds) Construction theory of functions of several variables. Springer, Berlin, pp 85–100
10. Farrokhabadi M, Browell J, Wang Y, Makonin W, Zareipour H (2021) Day-ahead electricity demand forecasting competition: Post-covid paradigm. Tech. rep.
11. Fasiolo M, Nedellec R, Goude Y, Wood SN (2020) Scalable visualization methods for modern generalized additive models. J Comput Graph Stat 29(1):78–86
12. Flaxman S, Mishra S, Gandy A, Unwin HJT, Mellan TA, Coupland H, Whittaker C, Zhu H, Berah T, Eaton JW, et al (2020) Estimating the effects of non-pharmaceutical interventions on COVID-19 in Europe. Nature 584(7820):257–261
13. Gaillard P, Goude Y (2015) Forecasting electricity consumption by aggregating experts; how to design a good set of experts. In: Modeling and stochastic learning for forecasting in high dimensions. Springer, pp 95–115
14. Gaucher S, Goude Y, Antoniadis A (2021) Hierarchical transfer learning with applications for electricity load forecasting. Preprint. arXiv:211108512
15. Goehry B, Yan H, Goude Y, Massart P, Poggi JM (2021) Random forests for time series. REVSTAT Stat J. https://hal.archives-ouvertes.fr/hal-03129751/
16. Hastie T, Tibshirani R (1990) Generalized additive models. Chapman & Hall

17. Hastie T, Tibshirani R (1993) Varying-coefficient models. J R Stat Soc Ser B Methodol 55(4):757–796
18. IEA (2020) Year-on-year change in weekly electricity demand, weather corrected, in selected countries. https://www.iea.org/data-and-statistics/charts/year-on-year-change-in-weekly-electricity-demand-weather-corrected-in-selected-countries-january-december-2020
19. Jones M, Pewsey A (2009) Sinh-arcsinh distributions. Biometrika 96(4):761–780. https://doi.org/10.1093/biomet/asp054
20. Liu Z, Ciais P, Deng Z, Lei R, Davis SJ, Feng S, Zheng B, Cui D, Dou X, Zhu B, Guo R, Ke P, Sun T, Lu C, He P, Wang Y, Yue X, Wang Y, Lei Y, Zhou H, Cai Z, Wu Y, Guo R, Han T, Xue J, Boucher O, Boucher E, Chevallier F, Tanaka K, Wei Y, Zhong H, Kang C, Zhang N, Chen B, Xi F, Liu M, Bréon FM, Lu Y, Zhang Q, Guan D, Gong P, Kammen DM, He K, Schellnhuber HJ (2020) Near-real-time monitoring of global co2 emissions reveals the effects of the covid-19 pandemic. Nat Commun 11(1):5172. https://doi.org/10.1038/s41467-020-18922-7
21. Mayr A, Fenske N, Hofner B, Kneib T, Schmid M (2012) Generalized additive models for location, scale and shape for high dimensional data—a flexible approach based on boosting. J R Stat Soc Ser C (Appl Stat) 61(3):403–427
22. Meyer BH, Prescott B, Sheng XS (2021) The impact of the covid-19 pandemic on business expectations. Int J Forecasting. https://doi.org/10.1016/j.ijforecast.2021.02.009, https://www.sciencedirect.com/science/article/pii/S0169207021000509
23. Molnar C (2019) Interpretable machine learning. Lulu.com
24. Pullano G, Valdano E, Scarpa N, Rubrichi S, Colizza V (2020) Evaluating the effect of demographic factors, socioeconomic factors, and risk aversion on mobility during the covid-19 epidemic in France under lockdown: a population-based study. Lancet Digital Health 2(12):e638–e649
25. Pullano G, Valdano E, Scarpa N, Rubrichi S, Colizza V (2020) Population mobility reductions during covid-19 epidemic in France under lockdown. MedRxiv 29:2020
26. Reeve HW, Brown G (2018) Diversity and degrees of freedom in regression ensembles. Neurocomputing 298:55–68
27. Rigby RA, Stasinopoulos DM (2005) Generalized additive models for location, scale and shape. J R Stat Soc Ser C (Appl Stat) 54(3):507–554. https://doi.org/10.1111/j.1467-9876.2005.00510.x
28. Wintenberger O (2017) Optimal learning with Bernstein online aggregation. Mach Learn 106(1):119–141
29. Wood SN (2006) Low-rank scale-invariant tensor product smooths for generalized additive mixed models. Biometrics 62(4):1025–1036
30. Wood SN (2013) On p-values for smooth components of an extended generalized additive model. Biometrika 100(1):221–228
31. Wood SN (2013) A simple test for random effects in regression models. Biometrika 100(4):1005–1010
32. Wood SN (2017) Generalized additive models: an introduction with R, 2nd edn. CRC Press, Boca Raton
33. Wood SN (2021) Inferring UK COVID-19 fatal infection trajectories from daily mortality data: were infections already in decline before the UK lockdowns? Biometrics. https://doi.org/10.1111/biom.13462, https://onlinelibrary.wiley.com/doi/full/10.1111/biom.13462
34. Wood SN, Pya N, Säfken B (2016) Smoothing parameter and model selection for general smooth models (with discussion). J Am Stat Assoc 111:1548–1575
35. Wood SN, Li Z, Shaddick G, Augustin NH (2017) Generalized additive models for gigadata: Modeling the uk black smoke network daily data. J Am Stat Assoc 112(519):1199–1210. https://doi.org/10.1080/01621459.2016.1195744
36. Ziel F (2021) Smoothed bernstein online aggregation for day-ahead electricity demand forecasting. Tech. rep., arXiv:2107.06268

Printed in the United States
by Baker & Taylor Publisher Services